The Palgrave Macmillan Animal Ethics Series

Series editors: **Andrew Linzey** and **Priscilla Cohn**

In recent years, there has been a growing interest in the ethics of our treatment of animals. Philosophers have led the way, and now a range of other scholars have followed, from historians to social scientists. From being a marginal issue, animals have become an emerging issue in ethics and in multidisciplinary inquiry. This series explores the challenges that Animal Ethics poses, both conceptually and practically, to traditional understandings of human-animal relations.

Specifically, the series will:

* provide a range of key introductory and advanced texts that map out ethical positions on animals;
* publish pioneering work written by new, as well as accomplished, scholars, and
* produce texts from a variety of disciplines that are multidisciplinary in character or have multidisciplinary relevance

Titles include

ANIMAL SUFFERING: PHILOSOPHY AND CULTURE
Elisa Aaltola

ANIMALS AND PUBLIC HEALTH:
Why Treating Animals Better Is Critical to Human Welfare
Aysha Akhtar

AN INTRODUCTION TO ANIMALS AND POLITICAL THEORY
Alasdair Cochrane

ANIMAL CRUELTY, ANTISOCIAL BEHAVIOUR AND AGGRESSION:
More than a Link
Eleonora Gullone

POWER, KNOWLEDGE, ANIMALS
Lisa Johnson

THE COSTS AND BENEFITS OF ANIMAL EXPERIMENTS
Andrew Knight

AN INTRODUCTION TO ANIMALS IN VISUAL CULTURE
Randy Malamud

POPULAR MEDIA AND ANIMALS
Claire Molloy

ANIMALS, EQUALITY AND DEMOCRACY
Siobhan O'Sullivan

AN INTRODUCTION TO ANIMALS AND SOCIOLOGY
Kay Peggs

SOCIAL WORK AND ANIMALS: A MORAL INTRODUCTION
Thomas Ryan

First published 2012 by
PALGRAVE MACMILLAN

Palgrave Macmillan in the UK is an imprint of Macmillan Publishers Limited, registered in England, company number 785998, of Houndmills, Basingstoke, Hampshire RG21 6XS.

Palgrave Macmillan in the US is a division of St Martin's Press LLC, 175 Fifth Avenue, New York, NY 10010.

Palgrave Macmillan is the global academic imprint of the above companies and has companies and representatives throughout the world.

Palgrave® and Macmillan® are registered trademarks in the United States, the United Kingdom, Europe and other countries

ISBN: 978–0–230–23923–4

This book is printed on paper suitable for recycling and made from fully managed and sustained forest sources. Logging, pulping and manufacturing processes are expected to conform to the environmental regulations of the country of origin.

A catalogue record for this book is available from the British Library.

A catalogue record for this book is available from the Library of Congress.

10 9 8 7 6 5 4 3 2 1
21 20 19 18 17 16 15 14 13 12

Printed and bound in Great Britain by
CPI Antony Rowe, Chippenham and Eastbourne

The Palgrave Macmillan Animal Ethics Series

Series editors: **Andrew Linzey** and **Priscilla Cohn**

In recent years, there has been a growing interest in the ethics of our treatment of animals. Philosophers have led the way, and now a range of other scholars have followed, from historians to social scientists. From being a marginal issue, animals have become an emerging issue in ethics and in multidisciplinary inquiry. This series explores the challenges that Animal Ethics poses, both conceptually and practically, to traditional understandings of human-animal relations.

Specifically, the series will:

- provide a range of key introductory and advanced texts that map out ethical positions on animals;
- publish pioneering work written by new, as well as accomplished, scholars, and
- produce texts from a variety of disciplines that are multidisciplinary in character or have multidisciplinary relevance

Titles include

ANIMAL SUFFERING: PHILOSOPHY AND CULTURE
Elisa Aaltola

ANIMALS AND PUBLIC HEALTH:
Why Treating Animals Better Is Critical to Human Welfare
Aysha Akhtar

AN INTRODUCTION TO ANIMALS AND POLITICAL THEORY
Alasdair Cochrane

ANIMAL CRUELTY, ANTISOCIAL BEHAVIOUR AND AGGRESSION:
More than a Link
Eleonora Gullone

POWER, KNOWLEDGE, ANIMALS
Lisa Johnson

THE COSTS AND BENEFITS OF ANIMAL EXPERIMENTS
Andrew Knight

AN INTRODUCTION TO ANIMALS IN VISUAL CULTURE
Randy Malamud

POPULAR MEDIA AND ANIMALS
Claire Molloy

ANIMALS, EQUALITY AND DEMOCRACY
Siobhan O'Sullivan

AN INTRODUCTION TO ANIMALS AND SOCIOLOGY
Kay Peggs

SOCIAL WORK AND ANIMALS: A MORAL INTRODUCTION
Thomas Ryan

AN INTRODUCTION TO ANIMALS AND THE LAW
Joan Schaffner

Forthcoming titles

HUMAN ANIMAL RELATIONS: THE OBLIGATION TO CARE
Mark Bernstein

ANIMALS IN THE CLASSICAL WORLD: ETHICAL PERCEPTIONS
Alastair Harden

The Palgrave Macmillan Animal Ethics Series
Series Standing Order ISBN 978–0–230–57686–5 Hardback
978–0–230–57687–2 Paperback
(*outside North America only*)

You can receive future titles in this series as they are published by placing a standing order. Please contact your bookseller or, in case of difficulty, write to us at the address below with your name and address, the title of the series and one of the ISBNs quoted above.

Customer Services Department, Macmillan Distribution Ltd, Houndmills, Basingstoke, Hampshire RG21 6XS, England

Animal Cruelty, Antisocial Behaviour and Aggression

More than a Link

Eleonora Gullone
Monash University, Australia

First published 2012 by
PALGRAVE MACMILLAN

Palgrave Macmillan in the UK is an imprint of Macmillan Publishers Limited,
registered in England, company number 785998, of Houndmills, Basingstoke,
Hampshire RG21 6XS.

Palgrave Macmillan in the US is a division of St Martin's Press LLC,
175 Fifth Avenue, New York, NY 10010.

Palgrave Macmillan is the global academic imprint of the above companies
and has companies and representatives throughout the world.

Palgrave® and Macmillan® are registered trademarks in the United States,
the United Kingdom, Europe and other countries

ISBN: 978–0–230–23923–4

This book is printed on paper suitable for recycling and made from fully
managed and sustained forest sources. Logging, pulping and manufacturing
processes are expected to conform to the environmental regulations of the
country of origin.

A catalogue record for this book is available from the British Library.

A catalogue record for this book is available from the Library of Congress.

10 9 8 7 6 5 4 3 2 1
21 20 19 18 17 16 15 14 13 12

Printed and bound in Great Britain by
CPI Antony Rowe, Chippenham and Eastbourne

Contents

List of Tables	viii
Foreword by Phil Arkow	ix
Series Editors' Preface	xvi
Acknowledgements	xviii

1 Introduction: The Aims of This Book — 1

2 Historical and Current Conceptualizations of Animal Cruelty — 5
Co-occurrence between animal cruelty and antisocial behaviour — 7
The evolution of current thinking — 8
The Child Behavior Checklist — 10
The Diagnostic and Statistical Manual for Mental Disorders — 10
Defining animal cruelty — 11
Animal cruelty motivations — 13
Chapter summary — 15

3 Conceptualizations of Antisocial Behaviour — 16
From behaviours to individuals — 17
From aggressive behaviours to antisocial behaviours — 17
Toward a more developmentally-oriented focus — 17
A developmental psychopathology approach — 18
Defining antisocial behaviour — 18
Defining aggression — 19
The importance of the intent requirement — 20
Aggression dimensions — 21
Chapter summary — 24

4 The Development of Antisocial Behaviour — 25
Key issues and considerations — 25
Development — 26
Infancy and toddlerhood — 26
Childhood — 27
Child-onset antisocial behaviour — 28
Adolescence — 30
Adolescent-onset antisocial behaviour — 30
Child versus adolescent-onset of antisocial behaviour — 31

Adulthood 33
Prevalence 34
Stability 35
Chapter summary 37

5 **Theoretical Accounts of Aggressive Behaviour
and Animal Cruelty** 39
Theoretical accounts of aggressive behaviour 39
Cognitive neo-association theory 40
Social cognitive models 40
General Aggression Model (GAM) 41
Theoretical accounts of animal cruelty 42
Antisocial behaviour risk factors 42
Chapter summary 43

6 **Biological and Individual Difference Risk Factors** 45
Biological risk factors 45
Sex differences 46
Baseline levels of arousal 47
Individual difference risk factors 48
Temperament 48
Personality 50
Psychopathy and Callous-Unemotional traits 51
Psychopathy in non-adult populations 54
Chapter summary 56

7 **Environmental Risk Factors** 58
Socioeconomic status 58
Provocation 59
Opportunity 59
Aggressive cues and exposure to violence 60
Family and parenting factors 62
Attachment relationships 64
Parenting practices 67
Parental warmth 67
Coercive and inconsistent parenting practices 68
Physical punishment 69
Direct and indirect abuse effects 71
Peer relationships 74
Chapter summary 75

8 **Emotional and Cognitive Processes** 78
Emotion processes 78
Emotion regulation 78
Cognitive factors and information processing 80

Knowledge structures	80
Schemas and scripts	81
Attributional, perception, and expectation biases	82
Accessibility of aggressive responses	83
Self-efficacy	84
Moral disengagement	85
Attitudes and beliefs	88
Chapter summary	89
9 Aetiological Accounts of Animal Cruelty	**91**
Theoretical models of animal cruelty	91
The Violence Graduation Hypothesis	91
The Deviance Generalization Hypothesis	98
Conduct disorder, antisocial personality disorder, and psychopathy	99
Criminal behaviour and animal cruelty	101
Federal Bureau of Investigation work	105
Family violence and animal cruelty	109
Bullying and animal cruelty in youth	111
Risk factors for the development of animal cruelty	113
Sex differences	115
Age differences	116
Witnessing of violence, and animal cruelty	117
Family and parenting experiences	119
Displacement of aggression	123
Cognitive errors, aggressive cues, and exposure to violence	124
Empathy and emotion regulation	125
Chapter summary	127
10 Conclusions and Future Directions	**129**
The development of animal cruelty behaviour	129
Cross-cultural research	130
An action agenda	134
Proposed strategies for change	135
Concluding comments	137
Chapter summary	137
References	140
Glossary	165
Index	175

Tables

1   Models of personality and their dimensions   52
2   Descriptions of the disorders associated with animal cruelty   102
3   Comparative percentages of crimes by category for all alleged offenders and only alleged animal cruelty offenders based on Victoria Police data for the years 1994 through 2001   104
4   Checklist of risk behaviours predictive of future violence   107

Foreword

How does violence, and particularly violence in childhood, begin? More importantly, how do we prevent it from happening? Philosophers, writers, and social scientists have asked these questions for thousands of years, and we are no closer today to ending violence than we were when Plato observed that without sufficient education, man can change from the most divine and most civilized to the most savage of earthly creatures.

The past four decades have witnessed a renaissance of interest in an age-old notion. This is the concept of what today we call "The Link," the idea that acts of interpersonal violence are frequently preceded by, or co-occur with, acts of cruelty to animals, "red flag" markers that previously were ignored. Animal abuse is becoming more widely recognized as a potential indicator and/or predictor of interpersonal violence that escalates in range, severity, and number of victims. This notion captivates popular thought with its intuitive appeal, but concomitantly challenges the research community in attempting to validate it empirically. Fortunately, a small, but rapidly growing, corpus of literature – as reviewed comprehensively in the book you are about to read – gives increasing credence to centuries of conventional wisdom that animal abuse and human violence often go hand-in-hand – or, as we sometimes quip, hand-in-paw.

But exploring the links among animal abuse, child maltreatment, intimate partner violence and elder abuse involves numerous methodological, semantic, cultural, and legal challenges, all anchored by an unfortunate reality: whether in the world of legislators, law enforcement officers, academics, or philanthropists, "animal" interests are perceived to be of lesser import than "human" interests. The ecological movement may have initiated wider recognition that human beings are, indeed, animals as well, and that the fates of all species are inextricably intertwined, but the general public and many professions still harbour internalized barriers that separate us (i.e., humans) from them (i.e., "beasts," "only animals," "the lower orders," and other terms about non-humans often applied derogatorily).

Some have called this "speciesism." Others attribute it to animals' lacking legal "standing" and their position, enshrined in law, as either chattel property (companion animals and livestock) or the property of the state (wildlife). In a court of law, a dog has the same status as a toaster. But no battered woman ever stayed in an abusive relationship to protect her toaster, whereas 18% to 45% of battered women report that fear for what the abuser will do to their pets prevents them from escaping sooner.

However society got here, the denigration of "animal" interests as less worthy than "human" needs is widespread and insidious. Veterinarians treat patients from multiple species who cannot tell them where it hurts, but consistently earn much less than their human medicine counter-parts. Animal shelters, if a community even has one, are often relegated to undesirable land next to a sewage treatment plant. Animal control officers are considered "dog catchers" rather than trained professionals enforcing laws with significant public health and safety implications for human, family, and community relations.

Among thousands of academic journals, only a handful are available for human-animal studies. Only 2% of charitable contributions in the U.S. is directed to animal causes; of that amount, some three-quarters is earmarked for environmental, wildlife, and conservation issues rather than companion animals. Of 76,000 American foundations, only 31 focus primarily on grants to animal protection issues. "Philanthropy," after all, means "love of people," and animal issues are not generally considered as germane to improving the human condition.

There are many of us who fervently believe that animals deserve a better break. Centuries of promoting kindness to animals as being the nice or right thing to do has not gotten us terribly far. Depicting animal abuse as a human welfare concern – as an under-recognized component of human aggression – is the way to make real progress among public policy makers, courts, and economic interests who have power to effect societal change. By positioning acts of animal abuse within the continuum of other antisocial behaviours, rather than as isolated incidents or acceptable childhood rites of passage, we can gain more progress not only in reducing animal abuse but also in improving human safety and lowering tolerance levels for all acts of aggression.

We live in a world with an unfortunate disconnect. On the one hand, society loves pets and respects them for giving us "unconditional love." But simultaneously, society considers animals peripheral to, and unworthy of, serious academic inquiry and public priority. This book is a major step forward in repairing that dichotomy. One hopes the reader will advance its message within various disciplines and break down the barriers that prevent social scientists, policy makers, and program administrators from considering the dark side of human-animal relationships as being worthy of their endeavours.

It is ironic that academia and other sectors have traditionally marginalized animal interests because the legal and moral underpinnings for animal welfare are rooted in human well-being. The United States has the world's oldest anti-cruelty laws, dating back to 1631 in our Colonial era. Yet, these statutes, similar to animal protection laws in other developed nations, have been enacted not because legislators perceive animals to have inherent "rights," but rather because acts of violence against animals are seen as harming another individual's property or as demeaning society's self-concept of what it means to be "civilized." Meanwhile, child protection agencies owe their origins to the animal protection movement.

Through organizations such as the National Link Coalition in the U.S.; similar groups in the U.K., Spain, and the Netherlands; and research and programs in Australia, New Zealand, Japan, Canada, South Africa, Ireland, the Bahamas, Italy, and elsewhere, professionals from multiple disciplines are coming together to address the nexus where various forms of family violence intersect. But researchers are caught in an uncomfortable "Catch-22" situation: Because animal cruelty is considered inconsequential, and animal issues are trivialized by everyone except the pet food and services industries, little funding is available, and The Link is not yet an accepted discipline because there is relatively little research. Ongoing cultural changes are needed to enable the study of animal cruelty to be more widely respected within the larger context of antisocial and aggressive behaviours.

In the criminal justice fields, animal welfare advocates are challenged by archaic and vague laws. Attempting to statutorily define universal standards of animals' welfare in terms of environmental conditions, nutritional needs, housing standards, objective criteria for the degree of animal pain and suffering experienced, and acceptable disciplinary practices and other human-imposed behaviours has proven devilishly difficult.

Meanwhile, legislators routinely exclude industry-accepted animal husbandry practices, culturally sanctioned activities, and even particular species from the purview of anti-cruelty laws. Combine these factors with the courts' needs to prove offenders' intentionality, the inability to objectively quantify "cruelty," and a marginal or minimal priority for these cases, and the result is a legal morass, inadequate or nonexistent enforcement, a woefully inadequate record of convictions, and reluctance by many courts to pursue prosecutions.

Even our language offers no relief. The semantics and symbolism of "cruelty" perplex legislators. "Cruelty" is enshrined in legal tradition, but many Link proponents prefer the term "abuse." The former implies malicious, purposeful intent that may be impossible to prove in a court of law; the latter follows the child abuse model utilized worldwide and suggests that maltreatment occurred regardless of the perpetrator's motivation. I submit that animal abuse is much like pornography: impossible to define, but you intuitively know it when you see it.

Whether the victims of family and community violence have two legs or four is often as much a matter of opportunity, misfortune, or accident as aggression, antisocial behaviour, or psychopathy. The origins of animal cruelty are now recognized as being multivariate, with often overlapping and contextual precursors, risk factors, and protective factors. They are as complex as the multiple explanations for interpersonal aggression. But however this abuse originates, society must recognize that animal abuse rarely occurs in a vacuum and is often part of a pattern of individual and/or familial dysfunction. We need to rebrand cruelty to animals as no longer an isolated act or as socially acceptable because "boys will be boys" or "it was only a cat." In so doing, there is much more likelihood that public policy makers, academic researchers, and law enforcement officials will give animal maltreatment the higher priority it deserves. By examining abuse along a continuum of severity marked by motivations and risk factors rather than by species, acts of violence against animals will be seen within the larger context of antisocial, aggressive behaviours.

The Link model breaks down the silos that have long separated the humane[1] and human services disciplines. It unites the public and non-government sectors, and bridges strata gaps between academia and social services agencies, in a common cause of violence prevention. It recognizes that we are all working with the same perpetrators and the same victims. It offers the potential for more comprehensive, multi-disciplinary, timely and effective prevention, assessment, intervention, and treatment proactivity and response.

Recognizing animal abuse as a potential indicator and predictor of human violence helps society to achieve numerous goals. Public policy

---

[1] In the U.S. and Canada, humane refers to the kindest qualities of an individual and is often associated with animal protection organizations. In the U.K., Australia, and several other Commonwealth nations, humane retains an older connotation associated with organizations that rescue drowning victims. The origin of these divergent meanings is unclear.

can be modified to protect all vulnerable members of the family and prevent further abuses. Interdisciplinary programs can achieve synergy and economies of scale through community collaborations and coalitions. Dynamic new research can transcend traditional boundaries and break new ground in understanding the aetiology of violence and describing strategies to confront it.

Fortunately, we are making steady progress, albeit slowly, at breaking down the barriers. It took until 2006 and the devastation of Hurricane Katrina, but the U.S. government now recognizes what we in the animal welfare movement had observed for decades: that the human-animal bond is so strong that survivors of disasters will not leave the rooftops of their inundated homes unless their pets can board the rescue helicopters with them. County governments must now have evacuation plans for pets as well as for humans to qualify for federal disaster relief funds.

As of this writing, 47 of our 50 American states have enacted laws that categorize certain acts of animal cruelty as felonies rather than as lower-prioritized misdemeanours or petty offenses. With rapidly increasing caseloads involving dogfighting, animal hoarding, and cruelty incidents, several cities now have prosecutors designated to handle highly specialized and challenging animal cruelty cases. The National District Attorneys Association has initiated a research and practice resource center, the National Center for Prosecution of Animal Abuse. Over 100 American schools of law now offer curricula in the specialty of animal law.

Twenty-two American states have recently enacted statutes enabling judges to include family animals within the purview of court-imposed protection orders issued in domestic violence cases. Six states include intimidating acts or threats of harm to animals within their statutory definitions of domestic violence or elder abuse. In some jurisdictions, cruelty to animals perpetrated in the presence of a child is considered a more serious crime with stronger sanctions.

Several states have enacted provisions to enable animal care and control officers, child abuse officials, veterinarians, and others in human services and law enforcement to cross-report suspected animal abuse and human violence to each other with immunity from civil or criminal liability. National veterinary associations in the United States, Canada, New Zealand, Denmark, and Norway endorse the concept of veterinarians not only treating the victims of animal cruelty but also reporting suspected cases to law enforcement authorities. Crime-scene investigation techniques, long employed in law enforcement, are being modified and developed as a specialized field of veterinary forensics to assist in the evidence-gathering, documentation and prosecution of animal cruelty cases.

More than 143 universities in the U.S., and 48 more in Australia, Canada, Germany, Great Britain, Israel, New Zealand, Poland, and Puerto Rico, offer courses and degree programs in human-animal studies. These curricula represent such interdisciplinary interests as philosophy, English, criminology, history, sociology, religion, anthropology, women's studies, social work, psychology, and counselling.

The health care field is beginning to take notice. Researchers are exploring the neurochemistry of human-animal interactions and the potential for pets to help fight obesity and motivate fitness through "Walk a Hound/Lose a Pound" exercise. Animal-assisted therapy is predicated on how human-animal interventions benefit human mental and physiologic health.

The "One Health" movement is bridging the commonalities between human and veterinary medicine. One Health proponents are collaboratively expanding their traditional linkages of zoonotic disease prevention, epidemiology, and food safety inspection to include a more holistic look at the human-animal bond. One Health transcends disciplinary and institutional boundaries to transform the way human and animal healthcare systems work together.

We do not – nor shall we ever – live in a perfect world. As Gullone observes so accurately in Chapter 10, "The core problematic issue relating to how best to deal with animal cruelty does not stem from a lack of prevention and intervention knowledge. Rather, the problem results from a lack of perceived worthiness of animal cruelty as a target for intervention. A change is needed in the perceived importance of animal cruelty among researchers, health care professionals and legislators. Such a change may slowly come about if animal cruelty becomes more strongly recognised and accepted as a characteristic behaviour of antisocial individuals or as a symptom of antisocial disorders such as Conduct Disorder or Psychopathy."

In an ideal world, of course, animal abuse should not have to be linked with human harm to be considered worthy of moral, legal, and academic consideration. But if that's what it takes, then that's what we'll do. Since history has shown that society addresses animal welfare concerns primarily when their impact on human well-being can be vividly demonstrated, The Link offers a tremendous opportunity. A Link perspective adds the weight of robust theoretical and empirical research to validate centuries of conventional wisdom to add clear, additional evidence for the strong connections between animal abuse and other behaviours that harm humans. Such research will not only advance the frontiers of knowledge but is desperately needed to change people's values about the importance of animals in our lives, and of animal cruelty as a valid marker of antisocial or aggressive individuals.

Phil Arkow
Stratford, N.J., U.S.A., March, 2012

Phil Arkow is the founder coordinator of the National Link Coalition, a multi-disciplinary collaborative network of individuals and organizations in human services and animal welfare who address the intersections among animal abuse, domestic violence, child maltreatment, and elder abuse through research, public policy, programming, and community awareness. He is a consultant to the American Society for the Prevention of Cruelty to Animals and the Animals & Society Institute. He recently directed Link Programs at the American Humane Association. He chairs the Latham Foundation's Animal Abuse and Family Violence Prevention Project and is an adjunct faculty member at Harcum College and Camden County College, where he teaches certificate courses in Animal-Assisted Therapy.

Series Editors' Preface

This is a new book series for a new field of inquiry: Animal Ethics.

In recent years, there has been a growing interest in the ethics of our treatment of animals. Philosophers have led the way, and now a range of other scholars have followed, from historians to social scientists. From being a marginal issue, animals have become an emerging issue in ethics and in multidisciplinary inquiry.

In addition, a rethink of the status of animals has been fuelled by a range of scientific investigations which have revealed the complexity of animal sentiency, cognition, and awareness. The ethical implications of this new knowledge have yet to be properly evaluated, but it is becoming clear that the old view that animals are mere things, tools, machines, or commodities cannot be sustained ethically.

But it is not only philosophy and science that are putting animals on the agenda. Increasingly, in Europe and the United States, animals are becoming a political issue as political parties vie for the "green" and "animal" vote. In turn, political scientists are beginning to look again at the history of political thought in relation to animals, and historians are beginning to revisit the political history of animal protection.

As animals grow as an issue of importance, so there have been more collaborative academic ventures leading to conference volumes, special journal issues, indeed new academic animal journals as well. Moreover, we have witnessed the growth of academic courses, as well as university posts, in Animal Ethics, Animal Welfare, Animal Rights, Animal Law, Animals and Philosophy, Human-Animal Studies, Critical Animal Studies, Animals and Society, Animals in Literature, Animals and Religion – tangible signs that a new academic discipline is emerging.

"Animal Ethics" is the new term for the academic exploration of the moral status of the non-human – an exploration that explicitly involves a focus on what we owe animals morally, and which also helps us to understand the influences – social, legal, cultural, religious and political – that legitimate animal abuse. This series explores the challenges that Animal Ethics poses, both conceptually and practically, to traditional understandings of human-animal relations.

The series is needed for three reasons: (i) to provide the texts that will service the new university courses on animals; (ii) to support the increasing number of students studying and academics researching in animal-related fields, and (iii) because there is currently no book series that is a focus for multidisciplinary research in the field.

Specifically, the series will

- provide a range of key introductory and advanced texts that map out ethical positions on animals;
- publish pioneering work written by new, as well as accomplished, scholars; and
- produce texts from a variety of disciplines that are multidisciplinary in character or have multidisciplinary relevance.

The new Palgrave Macmillan Series on Animal Ethics is the result of a unique partnership between Palgrave Macmillan and the Ferrater Mora Oxford Centre for Animal Ethics. The series is an integral part of the mission of the Centre to put animals on the intellectual agenda by facilitating academic research and publication. The series is also a natural complement to one of the Centre's other major projects, the Journal of Animal Ethics. The Centre is an independent "think tank" for the advancement of progressive thought about animals and is the first of its kind in the world. It aims to demonstrate rigorous intellectual enquiry and the highest standards of scholarship. It strives to be a world-class centre of academic excellence in its field.

We invite academics to visit the Centre's website www.oxfordanimalethics.com and to contact us with new book proposals for the series.

Andrew Linzey and Priscilla N. Cohn
General Editors

Acknowledgements

My motivation for writing this book was primarily to provide a voice for the many animals who suffer in silence and whose suffering often remains unseen and unacknowledged. Because they are powerless and cannot voice their suffering and victimization, non-human animals are easy victims. Crimes against them are not predictably punished, and when they are punished, the punishment rarely reflects the severity of the crime. It goes without saying that in human societies, non-human animals' suffering matters less. However, crimes against them reflect a violation of laws that are based on core human values. Such deviant behaviour is indicative of individuals who are disrespectful of the rights of others – human and animal. Thus, the suffering of non-human animals brought about through deliberate and callous behaviour should matter as much as that toward humans. Those who inflict animal cruelty are as dangerous to human well-being and survival as they are to non-human animals.

As this is the acknowledgements section of the book, it is appropriate that I indicate how my motivation for writing it evolved. As with most children, in my early years, I had a fascination with animals. I was lucky to grow up in a household where companion animals were a constant. My father Francesco Gullone enjoyed the innocence and playfulness of our companion animals, and this attraction to animals became a defining characteristic of my two sisters and me.

As a young child, I remember on more than one occasion, coming home and finding that my father had brought home an abandoned kitten. We had several cat companions throughout my growing years and spent many hours enjoying their company. When the neighbour's dog had puppies, we acquired one. My father named her Stella. She was an Australian Terrier, and as young girls we would place her in our doll's pram and wheel her around the backyard which backed onto a park with a river several metres away from our home. We spent most of our free hours with Stella in the park. She was a playmate, a friend, and also our guardian and protector. I remember on a number of occasions when my sisters or I dangled our feet into the river from the riverbank, she would grab onto our sweater sleeves and try to pull us back away from the river.

As an adult, I have continued to feel that a house is not a home without the comforting presence of companion animals. As documented in the literature on the benefits provided to humans by their companion animals, I find the presence of my companions soothing and warming. I often comment that petting my cats is like taking antistress medication. Their soothing purr works wonders, and my Pomeranian pooch keeps me active by demanding two walks per day.

My cats are now 17 years old. Peppy the pom is a mystery age as he was adopted from a shelter, but my guess is that he is nearing 15 years. I have shared my life with these three beautiful beings for a significant period of time, and my bond with them is strong. They have been there when I have celebrated joyful and momentous events in my life, and through some of the saddest moments in my life, such as my father's recent death. Their gentle presence and unconditional affection are predictable constants.

As I reflect on these life experiences, it is clear that my dear departed father was of pivotal influence on the importance I now place on animals in my life and on their welfare. I must acknowledge and thank my dear father for inspiring in me my affection for, and attraction to, animals. I must also thank my darling husband, Ajeet, for his encouragement and support and the many words of wisdom he imparted during the processes of planning and writing. Of significant importance, I must acknowledge and thank Professor Andrew Linzey for his suggestion and invitation that I write this book. Finally, I would like to acknowledge my two beautiful cats (Misty and Misky) and my pooch Peppy for their constant comforting presence whilst writing this book, and at many other times.

1
Introduction: The Aims of This Book

At the time of writing, there exist predominantly separate theoretical and empirical literatures on antisocial behaviour (incorporating aggression and violence) on one hand, and animal cruelty on the other hand. The primary aim of this book is to extend an empirically supported argument that animal cruelty behaviours fit logically and comfortably into existing theories of antisocial behaviour, thereby arguing for a harmonious marriage between these two separate literatures.

Supporting this marriage is the marked shift in the focus of theoretical and research attention that has occurred in the antisocial behaviour literature over the past few decades. Based on the finding that a select group of individuals is responsible for the majority of all crimes, including animal cruelty, a shift in research focus has occurred toward better understanding the development of chronically antisocial *individuals* rather than their *behaviour*, as was previously the case. Such a shift is consistent with the focus that has dominated animal cruelty research.

Current understanding of antisocial individuals is based on a large and sophisticated empirical base. Several chapters of the current work provide detailed reviews of the theoretical and empirical work that sheds light on the characteristics of antisocial individuals and the many factors that relate to the development of their behaviours. Research has indicated the importance of predispositional tendencies, environmental factors, and developmental histories of people who engage in aggressive or violent behaviours. This research is reviewed.

With regard to animal cruelty, the current work focuses on research and theory relating to individuals who engage in particular forms of animal cruelty, specifically those forms that are consistent with definitions of antisocial behaviour and human aggression. That is, the animal cruelty behaviours of focus include those performed with the

1

deliberate intention of causing harm (i.e., pain, suffering, distress and/ or death) to an animal. Included in this definition are both physical harm and psychological harm. Moreover, as detailed in the discussion on animal cruelty behaviours in Chapter 2, most published definitions of animal cruelty comprise a behavioural dimension including both acts of omission (e.g., neglect) and acts of commission (e.g., beating). There is general agreement that one important dimension of animal cruelty is indication that the behaviour occurred purposefully, that is, with deliberateness and without ignorance.

Thus, the focus of the current work is to examine those *individuals* who act with the *deliberate intention* to cause harm to non-human sentient beings, to examine their development and what motivates their behaviours. There are undoubtedly instances where there is instrumental gain that results from being cruel to animals or from engaging in other aggressive or antisocial behaviours. However, such instrumental gain is not a primary motivation of the behaviour.

It is acknowledged that animal cruelty is a much broader construct than that being focussed upon in the current work. As cogently stated by Peterson and Farrington (2007), "Defining animal cruelty is problematic, because of the existence of socially and culturally sanctioned activities which harm animals, differing attitudes toward different species, and a continuum of severity that can range from teasing to torture" (p. 23). Moreover, there are two broad perspectives on the issue.

According to the animal welfare position (e.g., Broom, 1991), although it is acceptable to use animals for human purposes including for clothing, food, research and entertainment, causing unnecessary and avoidable suffering constitutes cruelty. According to a second perspective known as the *animal rights perspective* (e.g., Regan, 1983), given the subjective nature of determining what is "unnecessary" suffering or "inhumane" treatment, the only way of being certain to avoid the suffering of animals brought about through human behaviour, is to avoid using animals for human ends altogether.

Of importance to this work is the level at which the animal cruelty is occurring. Not only are there intentional and unintentional cruelty behaviours, there are cruelty behaviours that apply at a national level, a societal level or an individual level. Generally, behaviours that apply at a national or societal level are culturally endorsed. That is, they are not socially unacceptable or deviant. The argument examined in the current work is that socially unacceptable animal cruelty behaviours are predominantly carried out by the same individuals who also behave in deviant ways toward other humans. That is, they are those individuals who behave in an antisocial manner. Thus, for example, tail docking in dogs, and rodeos are behaviours that do not fit within the current framework, since they are not considered deviant, at least at a national level. Despite such behaviours being consistent with an animal cruelty definition if one considers harming animals to be a sufficient criterion for defining animal cruelty, since they cannot be categorized as deviant or antisocial, they cannot validly differentiate between deviant and non-deviant or disordered and non-disordered individuals in the relevant society.

Thus, the different philosophical perspectives on the treatment of animals that apply to the broader animal cruelty literature will not be of focus. Nevertheless, it is relevant to the current work to consider how cultural and societal attitudes toward animals potentially compromise the "objectivity" of scientists. Prevailing and deeply ingrained attitudes within human cultures that consider using animals for human benefit to be acceptable inevitably lead to the position wherein animals' suffering is considered to be less worthy of both scientific and moral consideration (Linzey, 2009b; Rollins, 2006). This may, in part, explain why animal cruelty has predominantly been ignored when investigating antisocial individuals and their behaviour.

True objectivity, as espoused by the scientific community, is a worthy goal but one that is difficult, if not impossible, to attain. In his extensive writings on aggression and morality, Bandura (1990; 1999) emphasized the importance of cognitive processes that enable us to justify or rationalize our attitudes and behaviours, enabling potentially conflicting and opposing positions to be held by the same individual. These are discussed in Chapter 8. Scientific evidence itself points to the inevitable biases involved in empirical work. It logically follows that ingrained biases that cause the sentience and suffering of animals to be considered of lesser importance when compared to humans, will prevail in theory building and testing. It is not surprising then that, despite the evidence that deliberate animal cruelty is a form of antisocial behaviour perpetrated by those who engage in other antisocial behaviours, animal cruelty has continued to be treated predominantly as distinct from other forms of antisocial behaviour. Further, despite its predominantly "illegal" status, it is generally considered to be of significantly less importance compared to many other illegal behaviours. Consequently, it attracts significantly more lenient legal sanctions if any at all. Likewise, although the perpetrators of animal cruelty are predominantly the same individuals who engage in aggression and violence toward other humans, animal cruelty is given little attention by scientists and the judiciary alike. For example, although listed as a diagnostic criterion for Conduct Disorder, mainstream aggression researchers have largely tended to ignore animal cruelty, as have health care workers, law enforcement and legislative authorities.

It is my hope that this book will reinforce the efforts of those who have worked toward drawing attention to the inseparability of individuals who engage in socially unacceptable animal cruelty and those who engage in antisocial acts toward other humans, in the interests of protecting the victims – human and non-human.

Also of importance, but beyond the scope of this work, is research that examines larger questions such as investigation of the relationship between the types and degrees of animal cruelty that are sanctioned at a national and/or societal level and the prevalence of antisocial individuals in that particular nation or culture. The vast differences across cultures in the types of sanctioned animal cruelty are evident when one considers the existence of legalized kangaroo shooting in Australia (and the bludgeoning to death of in-pouch joeys); bear baiting in Pakistan; whale hunting in Japan; bear bile farming in China and Vietnam; and animal experimentation in the US, the UK, and Australia. As was argued in the 1500s, it is reasonable to expect that societies that sanction higher levels and degrees of animal cruelty will, as a nation and/or society, have lower general levels of empathy and compassion for all sentient beings, including humans.

It is also noteworthy that, despite cultural and species differences relating to the appropriate or inappropriate treatment of animals, there is general agreement across cultures regarding what is considered deviant or socially unacceptable behaviour toward other humans. As stated by the United Nations Human Rights Council: "Human rights are rights inherent to all human beings, whatever our nationality, place of residence, sex, national or ethnic origin, colour, religion, language, or any other status. We are all equally entitled to our human rights without discrimination. These rights are all interrelated, interdependent and indivisible" (United Nations – http://www.un.org/en/).

We know that there is cultural/national variation in the extent to which human rights are observed. We also know that animal rights are equally, if not more severely, compromised in nations where human rights are compromised, further reinforcing the relationship between the compassion of a nation and the compassion of its individual members toward sentience – both human and non-human. It follows that if we cultivate a culture of compassion toward our non-human citizens, current and future generations will benefit through reduced antisocial and violent behaviour toward all sentient beings.

2
Historical and Current Conceptualizations of Animal Cruelty

The belief that our treatment of animals is intimately associated with the way we treat human beings has a long history in philosophy as is documented in the works of classical writers including Pythagoras and Porphyry, medieval scholars including Thomas Aquinas, and early philosophers such as Montaigne (1533–1592) and John Locke (1632–1704) (Unti, 2008). Described as "one of the most influential thinkers of Europe and the late Enlightenment", Linzey (2009a) refers to the lectures on ethics given by Immanuel Kant between 1775 and 1780 in which he made reference to this relationship: "If he is not to stifle his human feelings, he must practice kindness towards animals, for he who is cruel to animals becomes hard in his dealings with men" (Linzey, 2009a; p. 1). Expressing similar sentiments, in his writings on cruelty and the education of children, philosopher John Locke, wrote:

> Children should from the beginning be bred up in an abhorrence of killing or tormenting any living creature, and be taught not to spoil or destroy any thing unless it be for the preservation or advantage of some other that is nobler... the custom of tormenting and killing other animals will, by degrees, harden their hearts even toward men; and they who delight in the suffering and destruction of inferior creatures, will not be apt to be very compassionate or benign to those of their own kind. (Locke, 1693/1989)

A number of other significant events in the following century were also reflective of the idea of a connection between animal cruelty and human aggression and violence. For example, in 1750, British artist William Hogarth produced his engraved series on *The Four Stages of Cruelty* depicting a progression from cruelty to animals to violence against humans. In 1890 in the U.K., in his Cruelty to Animals Bill, Lord Erskine wrote, "The abuse of that [human] dominion by cruel and oppressive treatment of such animals, is not only highly unjust and immoral, but most pernicious in its example, having an evident tendency to harden the heart against the natural feelings of humanity" (cited in Linzey, 2009a; p. 1). Events such as these reinforced the importance of promoting a humane ethic in children, for healthy psychological development. Such emphasis is also evident in the mandating of humane education in 1886, in the U.S. state of Massachusetts. By 1905, the states of Oklahoma and Pennsylvania had also passed compulsory education laws, and in 1909, the same occurred in Illinois (Unti, 2008). Today, in the U.S., at least 12 states have passed laws promoting or mandating that humane education be taught in the classroom.

In 1927, the Humane Education Society was established in the United Kingdom. Its website states:

> The founding members were passionate about helping people, especially the young and disadvantaged, develop a generous and compassionate character by a deeper understanding and compassion for animals, which would in turn benefit the whole of society. (Humane Education Society, UK; 2010)

During a lecture she read at *the 16th Annual Midcontinent Psychiatric Meeting,* anthropologist Margaret Mead (1964) described a number of case study reports of troubled adolescents whose behaviour included cruelty to animals. On the basis of such reports, Mead (1964) concluded that childhood cruelty to animals may be an important precursor to antisocial behaviour in adulthood and argued that:

> It would, therefore, seem wise to include a more carefully planned handling of behaviour toward living creatures in our school curriculum on the one hand, and to alert all child therapists to watch for any record of killing or torturing a living thing. It may well be that this could prove a diagnostic sign, and that such children, diagnosed early could be helped instead of being allowed to embark on a long career of episodic violence and murder. (p. 22)

Co-occurrence between Animal Cruelty and Antisocial Behaviour

It seems that at this time, cruelty to animals as an important sign of other problematic behaviour had caught people's attention. In 1961, in his book entitled *The Murderer and his Victim*, Macdonald proposed a triad of characteristics including enuresis, firesetting, and cruelty to animals that could potentially predict a child's propensity to crimes of aggression against humans in adulthood. In 1963, a study by Macdonald was published, which was based on 100 adults who had made threats to kill. On the basis of the findings, Macdonald concluded that amongst other factors, including extreme parental brutality, the triad is an unfavourable prognostic factor in those who threaten to commit homicide. In 1966, a study by Hellman and Blackman was published, in which they argued support for the triad on the basis of the finding that of 31 prisoners charged with aggressive crimes against people, as many as 74% (n = 23) had a history of the triad behaviours. Felthous and Bernard (1979) later replicated this finding as did others (Wax & Haddox, 1974a & b).

However, in more recent work, others (e.g., Felthous and Kellert, 1987; Heath, Hardesty, & Goldfine, 1984; Heller, Ehrlich, & Lester, 1984; Slavkin, 2001) have questioned the validity of the triad. For example, in their retrospective study involving the case reports of 1,935 violent and non-violent offenders, Heller, Ehrlich, & Lester (1984) investigated the incidence of cruelty to animals, firesetting, and enuresis. They found that the incidence of the triad did not differentiate between violent and non-violent criminals. However, cruelty to animals did occur with higher frequency among those charged with a violent crime compared to those charged with a non-violent crime.

In his more recent study, Slavkin (2001) examined the records for juveniles referred for firesetting behaviours. The records were for 78 fire-setters aged between 3 and 6 years, 240 aged between 7 and 10, 157 aged 11 to 14, and 413 aged 15 to 18 years. This resulted in a total sample of 888 fire-setters. They tested the hypothesis that the presence of enuresis, cruelty to animals, and firesetting would predict recidivistic firesetting. Although no relationship was found between fire-setting recidivism and enuresis, juveniles who were recorded as being cruel to animals were more likely than those who were not to engage in recidivistic firesetting behaviours. Thus, whilst doubt has been cast on the soundness of the triad, animal cruelty has been found to be a significant marker of other problem behaviours.

The evolution of current thinking

MacDonald (1979) wrote one of the earliest reviews of the literature examining the relationship between children and companion animals. He divided his review into two main areas of focus; the first reviewed research on the relationship between children and companion animals, and the second focused on animal cruelty. Amongst the positive and normative aspects of children's relationships with animals, MacDonald highlighted findings that have since been supported by more recent research (e.g., Melson, 2003), including the confidant role by companion animals for many children, the comfort and emotional support provided by companion animals at difficult times, and their role as a family member. He also referred to the much-cited work of Levinson (1972) demonstrating the valuable role played by companion animals in "pet-facilitated psychotherapy".

MacDonald's discussion of animal cruelty highlighted the lack of available scientific investigation at that time whilst also acknowledging the recognition of its importance. He further stated, "Even a cursory examination of the customs of different countries or of different periods of history, reveals the importance of social and cultural factors in relation to ill-treatment of animals" (p. 352). MacDonald quoted from the works of his contemporaries including Ryder (1973) and Anthony (1973 cited in MacDonald, 1979), who suggested that children may pass through a developmental period during which cruelty to animals presents as a part of normal developmental processes involving experimentation with, and development of, a sense of control over one's environment. In addition to this thinking, other writers of the time, whose ideas concurred with those expressed by Mead, were also referred to by MacDonald. These included Gordon (1939), who argued that cruelty to animals was a special form of destructive behaviour, usually learned from parents. Brittain (1970 cited in MacDonald, 1979) expressed ideas similar to those expressed by current authors (e.g., Ascione, Thompson, & Black, 1997; Kellert & Felthous, 1985), that animal cruelty can be motivated by a desire for power over others and that this represents a driving motivation, over and above the infliction of pain.

On the basis of his review, MacDonald arrived at a conclusion which is incredibly insightful given current understanding of animal cruelty. He stated:

> There are a number of features associated with cruelty to animals. It is believed to occur in a minor form in many children as a

developmental stage. It is related to the concept of dominance and to violence, particularly murder. As a presenting symptom in child psychiatry it is seen mainly by boys of school age, and is linked with brain damage or mental retardation. Case studies emphasize unstable parenting with gross parental neglect, brutality or rejection. Alone or in combination with firesetting and enuresis it is believed by some authorities to be predictive of later violent behaviour. Isolated episodes may be developmental, but persistent or recurrent cruelty to animals seems to have value as a warning sign of later violence. (pp. 354–355)

Around three decades later, following in excess of 100 empirical studies examining cruelty to animals, there is agreement with MacDonald's conclusions on a number of grounds. First, mild and isolated forms of animal cruelty may be part of normal exploratory developmental behaviour (Ascione & Lockwood, 2001; Ascione, Thompson, & Black, 1997). Second, as is reviewed in detail in Chapter 9, in current thinking, there is no question that parenting experiences play an important aetiological role (e.g., Repetti, Taylor and Seeman, 2002) in the aetiology of aggression displayed by children. Third, as is detailed further below, it is generally agreed that animal cruelty is one of a number of behaviours listed as important for the prediction of later antisocial (including violent) behaviour (American Psychiatric Association, 2000; Gelhorn, Sakai, Price, & Crowley, 2007). Finally, there is strong consensus that persistent or recurrent cruelty to animals, as compared to isolated episodes, has important predictive value (e.g., Felthous & Kellert, 1987; Hensley, Tallichet, & Dutkiewicz, 2009).

The dichotomy of focus in animal research reflected in MacDonald's review continues to run through more recent human-animal relations literature. The first major focus involves the positive aspects of the relationship between people, particularly children and their companion animals (e.g., Bowd & Bowd, 1989; Bryant, 1990; Melson, 1988; 1990; Olsen, Quigley, & Stryker, 1988; Poresky, Hendrix, Mosier, & Samuelson, 1988; Poresky & Hendrix, 1990). The second major focus is on the negative aspects of interactions between humans and animals, and these predominantly involve animal abuse or cruelty. In particular, as reviewed in detail in Chapter 9, attention has been given to the relationship between human aggression or violence and animal cruelty, otherwise referred to as "the link" (e.g., Arluke, Levin, Luke, & Ascione, 1999; Ascione, 1993; 1999; 2001; Felthous, 1980; Felthous & Kellert, 1986; 1987; Kellert & Felthous, 1985; Volant, Johnson, Gullone, & Coleman, 2008).

The Child Behavior Checklist

Developments in the social sciences and criminology that have significantly influenced professional thinking and scientific writing on animal cruelty include the 1978 publication of the Child Behaviour Checklist (CBCL) as a research and clinical tool for the assessment of children's competencies and behavioural problems (Achenbach, 1978). The development of the CBCL was based on a survey of existing literature and the case histories of 1,000 child psychiatric patients conducted by Achenbach (1966). The CBCL continues to be one of the most widely used measures in child psychology and psychiatry. It is noteworthy that the checklist comprises a number of sub-scales including a cruelty sub-scale. Items in this scale include "Attacks people", "Fighting", "Cruel to others", "Cruel to animals" (Achenbach & Edelbrock, 1979).

The Diagnostic and Statistical Manual for Mental Disorders

The inclusion of cruelty to animals as a symptom of Conduct Disorder in the third edition of the Diagnostic and Statistical Manual of Mental Disorders (DSM-III; American Psychiatric Association, 1987) was likely, at least in part, to be influenced by the item content of the CBCL. This recognition undoubtedly played an important part in legitimizing animal cruelty as a behaviour worthy of scientific inquiry in relation to mental health.

Conduct Disorder is characterized by "a repetitive and persistent pattern of behaviour in which the basic rights of others or major age-appropriate societal norms or rules are violated" (Diagnostic and Statistical Manual-IV – Text Revised (DSM-IV-TR; American Psychiatric Association, 2000, p. 98). Conduct Disorder comprises a cluster of externalizing behaviours, including aggression, lying, stealing, destructiveness, cruelty (to humans and animals), and running away. Many of the behaviours associated with Conduct Disorder may occur, in varying degrees, in most children over the course of normal development (Kazdin, 1987). These behaviours are clinically important if "there is significant impairment in everyday functioning at home or school, or when behaviors are regarded as unmanageable by significant others (i.e., major caregivers such as parents and teachers)" (Kazdin, 1987; p. 187).

Conduct Disorder along with Antisocial Personality Disorder is among a cluster of disorders classified along what is referred to as the "Externalizing Spectrum". This spectrum encompasses personality traits that are disinhibitory in nature, such as impulsivity and aggression (Tackett & Krueger, 2011).

Compared to the third editions (DSM-III, and DSM-III Revised) of the Diagnostic and Statistical Manual of Mental Disorders, the fourth edition, published in 1994, reflected a broader understanding of cruelty to animals. In that edition, cruelty to animals was listed under the heading "Aggression toward People and Animals" along with other symptoms of violence against others rather than against property. As noted by Unti (2008), this reflected thinking that animals are not simply property and that underlying motivations for animal cruelty are likely to be different from those associated with vandalism or property damage. There is an implicit recognition that animal cruelty involves harm or death to sentient creatures.

Thus, from as early as the 1600s (as reflected in Locke's thinking), attention was being given to the development of violent and aggressive behaviours in relation to cruelty and abuse of non-human animals. However, the incorporation of animal cruelty into the broader aggression literature was not maintained. Research and theoretical attention on animal cruelty evolved into a largely separate literature, as is evident from the discussion that follows regarding the conceptualization of animal cruelty.

Defining animal cruelty

Although the terms "cruelty" and "abuse" have been used interchangeably by some writers, it is noteworthy that distinctions have been made. According to Peterson and Farrington (2007), cruelty behaviour results in some form of gratification to the offender for the suffering caused. In contrast, abuse has been referred to as the use of an inappropriate level of force (Peterson & Farrington, 2007). It is also noteworthy that perceptions about appropriate and inappropriate or humane and inhumane treatment of animals vary both historically and culturally.

Among the most often cited definitions of animal cruelty is that put forth by Frank Ascione, who defined it as "socially unacceptable behaviour that intentionally causes unnecessary pain, suffering, or distress to and/or the death of an animal" (Ascione, 1999; p. 51). Others, such as that by Felthous and Kellert (1986), define *substantial cruelty to animals* as a behaviour pattern that deliberately, repeatedly, and unnecessarily causes hurt to vertebrate animals in such a way that is likely to cause them serious injury. Brown (1988) defined cruelty as "unnecessary suffering knowingly inflicted on a sentient being (animal or human)". Brown made clear in his definition that the suffering may be of a physical type as in causing the sensation of pain, or it may be suffering that causes distress or psychological hurt such as would be the case with maternal deprivation. Brown also argued that cruelty to animals can be both positive or negative such that committing an act against the animal would constitute a positive form of cruelty, whereas failing to act as in neglecting to feed an animal or to care for it appropriately would constitute a negative form of cruelty.

Following detailed consideration of a number of definitions of animal cruelty, Dadds, Turner, and McAloon (2002) noted that most definitions comprise a number of features. These include a behavioural dimension that can be in the form of acts of omission (e.g., neglect) or acts of commission (e.g., beating). Another key characteristic is indication that the behaviour occurred purposely, that is, with deliberateness and without ignorance. Another definitional criterion is that the behaviour can bring about physical and/or psychological harm. Incorporating these definitional criteria, Dadds (2008) defined animal cruelty as a repetitive and proactive behaviour (or pattern of behaviour) intended to cause harm to sentient creatures.

As stated in Chapter 1, given that the emphasis in the current work is on animal cruelty behaviours that fit into the spectrum of antisocial behaviours, the focus will necessarily be those that have been defined by society and the legal system as socially unacceptable (c.f., Ascione's, 1999 definition). Unfortunately, many other behaviours that cause suffering to animals (e.g., hunting, farming, animal experimentation, behaviours involved in specific cultural or religious rituals) are socially and legally sanctioned. Given that they are socially acceptable and therefore non-deviant, they cannot meaningfully distinguish between deviant and non-deviant or pathological and non-pathological individuals.

Taking into account the specific aims of this work, the definition of animal cruelty that will be adopted herein is as follows:

Animal cruelty is behaviour performed repetitively and proactively by an individual with the deliberate intention of causing harm (i.e., pain, suffering, distress and/or death) to an animal with the understanding that the animal is motivated to avoid that harm. Included in this definition are both physical harm and psychological harm. As per the literature on human aggression, animal cruelty at the more extreme end of the aggression dimension (e.g., burning whilst alive, torture – c.f., murder, rape, assault as compared to teasing, hitting, tormenting), should be considered to be a violent sub-type of animal cruelty.

Certainly, more consideration needs to be given to the severity of acts of animal cruelty than is currently the case. In this regard, research into the motivations underlying animal cruelty is most useful.

Animal cruelty motivations

A number of authors (e.g., Dadds et al., 2002; Hensley & Tallichet, 2005; Merz-Perez, Heide, & Silverman, 2001) have emphasized the importance of determining the motivations underlying animal cruelty in order to better understand the processes underlying the behaviour, and particularly relationships with human violence and aggression. To this end, Kellert and Felthous' (1985) nine motivations have been referred to. They include (i) attempts to control an animal (e.g., hitting a dog to stop it barking), (ii) retaliation (e.g., use of extreme punishment for a perceived transgression on the part of the animal such as throwing a cat against a wall for vomiting in the house), (iii) acting out of prejudice against a particular species or breed. Such a motivation is accompanied by the belief that the particular animal is not worthy of moral consideration (cf. Bandura's, moral disengagement theory reviewed in Chapter 8), (iv) the expression of aggression through an animal (e.g., organizing dog fights), (v) the motivation to enhance one's own aggression (e.g., using animals for target practice or to impress others), (vi) shocking people for amusement (cruelty that is very overt and observed by others), (vii) retaliating against another person or as revenge (e.g., killing or maiming the companion animal of a disliked neighbour), (viii) displacement of aggression from a person to an animal. This motivation typically involves frustrated aggression. Many of the aggressive participants in Kellert and Felthous' (1985) study reported being physically abused as children. Participants' self-reports were supportive of displaced aggression, typically involving authority figures whom they reported hating or whom they feared so much that it prevented them from expressing their aggression directly. Their cruelty toward animals reportedly served as a displaced expression of the violence they experienced. As stated by Kellert and Felthous, "It is often easier in childhood to be violent toward an animal than against a parent, sibling, or adult" (p. 1124), and finally (ix) "non-specific sadism" refers to the desire to inflict suffering, injury or death in the absence of any particular hostile feelings toward an animal. One primary goal expressed within this motivation was to derive pleasure from causing the suffering. This motive was explained by Kellert and Felthous as sometimes being related to a desire to exercise power and control over an animal as a way of compensating for feelings of weakness or vulnerability.

In their 2005 study, Hensley and Tallichet examined demographic and situational influences on motivations to abuse animals. Their sample comprised 261 inmates from both medium- and maximum-security prisons. Their results revealed that anger was a motivation reported by nearly half of the sample. Nearly one-third of the sample reported that they were cruel to animals for fun. The researchers argued that both motivations are emotional responses on opposite ends of one continuum. Inmates who were cruel to animals when alone were more likely to commit the acts out of anger. In contrast, those inmates who were cruel in the presence of others indicated the need to impress or imitate those others. Respondents who reported covering up their acts of animal cruelty were around four times more likely to have engaged in bestiality. Another noteworthy finding was that the inmates who reported engaging in multiple acts of animal cruelty were around three times more likely to be motivated by a desire to control the animals when compared to those who engaged in fewer acts.

It is noteworthy that although Kellert and Felthous proposed a nine-motivation typology, their data highlighted that the animal cruelty was multidimensional, and one motivation alone was rarely reported by the participants in the study. Hensley and Tallichet (2005) also found that the same act of animal cruelty may have multiple motivations. It is also noteworthy that being motivated by anger was prominent as was the reporting of a sense of gratification from causing suffering and from deriving a sense of control over the animal. These motivations overlap somewhat with behaviours motivating individuals who display Callous-Unemotional traits. (See Chapter 6 for a detailed discussion of these traits.) According to Hensley and Tallichet (2005), cruelty motivated by the desire to control animals could be considered a form of displaced aggression.

Moreover, the finding of multiple motivations is consistent with conceptualizations of human aggression, as will be discussed in the subsequent chapter. For example, Anderson and Huesmann (2003) have argued that it is useful to characterize aggressive behaviour as varying along a number of identified dimensions (e.g., hostile affect versus agitated affect; whether the primary or ultimate goal is to cause harm to the victim versus an instrumental goal of deriving a profit or reward), thereby acknowledging its multidimensional nature.

Chapter Summary

In this chapter, a historical overview is provided regarding conceptualizations of animal cruelty. The perceived link between animal cruelty and antisocial behaviour toward humans dating back to the 1500s is highlighted. Since that time, a number of prominent individuals, including philosopher John Locke and anthropologist Margaret Mead, have echoed the view that cruelty to animals hardens the heart and desensitizes one to the suffering of others. Such a view has been credited with influencing the mandating of humane education in at least 12 U.S. states. In the U.K., this thinking is reflected in the establishment of an Humane Education Society in the early 1920s.

By the late 1900s, scientific thinking along these lines strengthened, based largely on systematic data showing a relationship between cruelty to animals and cruelty to humans (Achenbach, 1978). The inclusion of "being cruel to animals" as a symptom of Conduct Disorder diagnosis in the 1987 version of the *Diagnostic and Statistical Manual of Mental Disorders* published by the American Psychiatric Association was perhaps the single most influential factor spearheading a consistent flow of research examining this link.

Determining an agreed-upon definition of animal cruelty has also helped to progress research in the area. The main criteria comprising such a definition are provided, and the conceptual overlap with definitions of human aggression are noted. What follows is a discussion of the documented motivations underlying animal cruelty. Here, too, what is noteworthy is that, as with human aggression, animal cruelty has rarely been documented to be explained by one motivation but rather has been found to be multidimensional in nature.

As will be highlighted in subsequent chapters, there is substantial overlap in the way in which animal cruelty is conceptualized and the way in which antisocial behaviour and aggression are conceptualized. The subject of the next chapter is the conceptualization of the antisocial behaviour and aggressive behaviour.

3
Conceptualizations of Antisocial Behaviour

It is generally accepted and understood that antisocial behaviours which include aggression and violence are disruptive, not only to the individual but also to communities and society as a whole (Dishion, French, & Patterson, 1995). Of note, antisocial behaviours are highly stable over time such that childhood antisocial behaviour is predictive of adult antisocial behaviour. Furthermore, having a history of antisocial behaviour is predictive of a large range of problems during adulthood, including criminal behaviour, work failure, and troubled marriages. Also of concern is the intergenerational transfer of aggressive and antisocial behaviour (Serbin & Karp, 2004). This is not surprising given that, as parents, antisocial adults model and elicit antisocial behaviours in children (Patterson, Dishion, & Bank, 1984). At the more extreme end of the antisocial behaviours continuum is violence, which has been reported to be one of the leading public health problems worldwide with over 1.6 million lives lost each year and countless more being harmed (Krug, Dahlberg, Mercy, Zwi, & Lozano, 2002).

In their recent review of research and conceptualization of aggression and antisocial behaviour in youth, Dodge, Coie, and Lynam (2006) noted that over the past 40 years, crime rates have steadily risen in all countries that keep reasonably accurate records. Of particular concern, violent crimes committed by young juveniles have increased. Anderson and Bushman (2002) also note that since World War II, across all industrialized countries, homicide rates have increased. Proposed causes of increased violence include access to guns, global warming, violence against children at home and at school, as well as widespread exposure to violent entertainment media.

From behaviours to individuals

As a consequence of increasing crime rates, there has been a marked shift in the focus of theoretical and research attention to better understanding the development of chronically antisocial individuals (Hartup, 2005). Prior to the 1980s, the study of aggression focused more on the analysis of the behaviour and its development (e.g., Bandura, 1978), whereas in present times there has been a marked shift from focus on the *aggressive* act to focus on the development of aggressive *individuals*. The importance of this focus is clear when one considers the documented findings that chronically violent youth, who constitute a significant but relatively small group, are responsible for more than half of all crimes (Howell, Krisberg, & Jones, 1995).

From aggressive behaviours to antisocial behaviours

Over the past decade, it has also become increasingly clear that aggressive behaviours mostly occur within the context of other antisocial behaviours, including lying, stealing, destruction of property, burglary, sexual assault, and other violent crimes (Hartup, 2005). Given the co-occurrence between aggressive behaviour, most notably physical aggression with other forms of antisocial behaviour such as illicit drug use, it has been determined necessary to broaden the focus of research in the area and include aggression within the broader class of antisocial behaviours (Dodge et al., 2006). Pertinently, as stated by Dishion et al. (1995), "the frequency and variety of antisocial acts are the best predictors of more serious forms of antisocial behaviour, including violence" (p. 422). In support of their statement, Dishion and colleagues cite the findings of longitudinal work by Farrington (1991) showing that individuals who committed a variety of antisocial acts throughout their childhood and adolescent years were more likely to escalate to more serious violent acts. The findings also showed that those who were the more frequent offenders were more likely to be involved in violent offences as determined by their arrest record. Moreover, not only is aggressive behaviour relatively stable as an individual characteristic, stable aggression is a good predictor of antisocial behaviour during adolescence and adulthood (Moffitt, Caspi, Rutter, & Silva, 2001).

Toward a more developmentally-oriented focus

In an attempt to better understand the developmental trajectories of individuals who can be classified as antisocial, another major trend has been a shift toward more developmentally-oriented studies of aggression (Hartup, 2005). Earlier work in this area largely neglected a developmental focus. When age was examined in research such as in normative studies, only a limited age span (e.g., early childhood) tended to be encompassed. In contrast, more recent work has taken a broader developmental approach by examining continuities and discontinuities across the lifespan with many studies adopting longitudinal designs. For example, studies have charted criminal careers and found that they typically begin in childhood (Blumstein & Cohen, 1987).

A developmental psychopathology approach

The shift in focus also reflects the thinking that developmental trajectories and risk factors for pathologically antisocial individuals deviate from those for individuals showing normative trajectories. This is consistent with a developmental psychopathology perspective. Developmental psychopathology constitutes the application of a developmental perspective to the ever-changing relationship between adaptive and maladaptive outcomes, between pathology and normality, and between intrinsic and extrinsic influences on ontogenesis (Cicchetti, & Cohen, 1995). The incorporation of developmental psychopathology principles into the conceptualization of antisocial behaviour has been of particular significance for better understanding antecedent factors and pathways. In this regard, developmental psychopathologists are particularly interested in individuals at high risk for the development of disorders but who do not develop disorders. The understanding of how children overcome significant risk factors and proceed on an adaptive developmental course is invaluable for preventative and treatment avenues (Cicchetti & Crick, 2009; Cicchetti & Curtis, 2007). A developmental psychopathology perspective is also useful for defining different causal pathways that may lead to antisocial behaviour by uncovering factors that may disrupt, or that differ from, the normative developmental trajectory.

Defining antisocial behaviour

As noted by Rutter (2003), the field of antisocial behaviour has been criticized as lacking adequate conceptualization of the antisocial construct. This is reflected in the research literature which predominantly reports on studies of more homogenous constructs (e.g., aggression, Conduct Disorder, bullying) rather than the broad construct of antisocial behaviour. Rutter goes on to state that an adequate conceptualization of the construct must encompass a large range of socially disapproved behaviours "if only because the evidence makes clear that few criminals are highly specialised in the crimes that they commit" (p. 374). He then goes on to consider possible characteristics that would provide some cohesion across the range of behaviours encompassed in the construct. Included are considerations of overt physical aggression, lack of concern for the feelings of others, sensation-seeking or risk-taking behaviours, and non-conformity or disregard for the rules of society or social institutions. However, whilst each is certainly a characteristic of certain antisocial behaviours, none is a characteristic of them all. At the end of his deliberations, Rutter comes no closer to a cohesive conceptualization of the construct.

One way of dealing with this issue as adopted by Dodge and colleagues (2006) is not only to acknowledge the conceptual and empirical overlap between antisocial behaviour and aggressive behaviour but to also identify the differences. In doing so, these authors note that aggression has been defined as behaviour which has the intention of harming or injuring another person or other people. In contrast, in recognizing its broader reference, Loeber (1985) has defined antisocial behaviour as that which causes mental or physical harm, property loss, or damage to others. Frick and Viding (2009) provide a somewhat more refined definition of antisocial behaviour as "criminal and aggressive behaviors that violate the rights of others or major societal norms" (p. 1111). Given the heterogeneity of behaviours subsumed within antisocial behaviour, attempts have been made to identify meaningful subtypes of antisocial individuals depending upon the patterns of antisocial behaviour they exhibit (e.g., Burt, 2012). Developmentally, an additional focus has been to determine whether youth can be classified depending upon the trajectory of their antisocial behaviour over time. Examples provided by Frick and Viding (2009) include whether the behaviour displayed is stable, increasing, or decreasing in occurrence over time.

As with defining antisocial behaviour, defining aggression has involved several issues of contention.

Defining aggression

Given the evolved focus from aggressive acts to individual differences in aggressive behaviour, Dodge et al. (2006), argue that the need for precision in defining aggressive *acts* may at the present time be of less significance than the need for better understanding and identification of aggressive and antisocial *individuals*. Nonetheless, achieving some consensus on what constitutes aggressive behaviour remains important. To this end, Dodge et al. (2006) argue that aggression be defined as behaviour that aims to harm or injure another or others. Others have similarly defined aggression. For example, Anderson (2002) has defined aggression as behaviour performed by a person (the aggressor) with the immediate intention of harming another person (the victim). The perpetrator (aggressor) must believe that the behaviour will harm the victim and that the victim is motivated to avoid that intended harm. Whilst Gendreau and Archer (2005) have argued that *harm and injury* to others are the strongest indicators that an aggressive act has occurred, it is noteworthy that, in his definition, Anderson (2002) states that *actual* harm is not a requirement. Moreover, although research has predominantly focussed on physical harm, mental or psychological harm are included as consequences of aggression as is the case, for example, in persistent bullying. Consistently, Anderson's definition includes "harm" that can be either or both physical harm (e.g., punching someone) and psychological harm (e.g., verbal abuse). Indirect harm (e.g., destroying the victim's property) can also constitute aggression (Anderson, 2002). It is worth noting the consistencies between the definitions of aggression and animal cruelty.

The importance of the intent requirement

The importance of the "intent" requirement though included by a number of authors in their definition of aggression (e.g., Anderson & Huesmann, 2003; Dishion et al., 1995), and considered to be the differentiating feature between aggression and antisocial behaviour (Dodge et al., 2006), has been considered problematic (Loeber & Hay, 1997). As stated by Dishion et al. (1995), "The obvious difficulty with intention as a definitional component of antisocial behaviour is the black box problem. How do we know if an act was intentional?" (p. 421). Intent to harm is difficult to prove and easy to deny (Loeber & Hay, 1997). It is also a problematic criterion when investigating aggression in children (Gendreau & Archer, 2005), and problematic in cases of indirect aggression where the perpetrator is motivated to conceal his or her actions. When considering this problem, Dodge et al. (2006) argue that it is not a definitional one but a measurement problem, and further state that "To confound the conceptual definition with the criterion for measurement is to fall prey to the limits of operationism – a philosophical approach to science that guided much of early experimental psychology" (p. 722). Other issues involved in defining aggression relate to whether it should be conceptualized into different types or whether it should be seen as one behaviour varying along a number of dimensions (e.g., Anderson & Huesmann, 2003; Gendreau and Archer, 2005).

Aggression dimensions

The following section will discuss the main dimensions upon which aggression has been classified or dichotomized. These include dimensions of affective (hostile) versus instrumental, proactive versus reactive and impulsive versus premeditated aggression. Other differences proposed in aggressive behaviours depend upon whether the aggression is physical, verbal, direct or indirect. As noted by Anderson and Huesmann (2003), whilst the difference between verbally and physically aggressive behaviours is obvious, the direct versus indirect distinction is less so. Specifically, whilst indirect aggression occurs outside the presence of the intended target, and includes behaviours such as taking a person's possessions or telling lies about them in their absence, direct aggression occurs in the presence of the target. Of particular relevance to physical aggression, is the dimension of severity. As stated by Anderson and Huesmann (2003), violence is physical aggression that lies at the extreme high end of the aggression continuum and includes murder, rape, and aggravated assault. In distinguishing between aggression and violence, Anderson and Huesmann note that whilst only extreme aggression can constitute violence, all violence is aggressive.

Gendreau and Archer (2005) cited an important dichotomy in aggression based on the work of Feshbach (1964) which distinguishes between behaviours that have the primary goal of causing injury to the victim and pleasure or satisfaction to the aggressor compared to behaviours that do not have injury as the main goal. In the latter case, the behaviour is motivated by the primary goal of obtaining a reward. This type of aggression has been referred to as *instrumental aggression*. In contrast, the former subtype of aggression is referred to as *hostile aggression*. Anderson and Huesmann (2003) have noted that hostile aggression is also referred to as *affective* or *emotional aggression* and is characterized as being impulsive, unplanned, and thoughtless. It is predominantly motivated by anger and tends to occur in response to perceived provocation on the part of the aggressor. It is a proactive, rather than reactive, form of aggression and has been described as resulting from cold calculation rather than hot affect.

It is noteworthy that Kellert and Felthous' (1985) nine-motivation typology for animal cruelty has some consistencies with the various proposed dimensions of human aggression. For example, the motivation to aggress against an animal to impress others can otherwise be considered to be an instrumental form of aggression. The behaviour that Kellert and Felthous (1985) labelled as non-specific sadism which is motivated by the goal of deriving pleasure from an animal's suffering can be classified as hostile aggression according to Anderson and Huesmann's (2003) conceptualization.

Gendreau and Archer (2005) have also differentiated aggression, depending upon whether it is offensive (proactive) or defensive (reactive), and on the basis of the motivating emotion (e.g., anger or fear). Aggression that constitutes a response to some form of provocation has been referred to as *reactive aggression*. This type of aggression has been described as being an impulsive response to a threat or provocation (Gendreau & Archer, 2005). In contrast, *proactive aggression* is characterized by forethought. It is not as readily associated with a proximal elicitor and is more controlled and premeditated. It is also less emotionally reactive. Anderson and Huesmann (2003) have described reactive and proactive aggression in similar terms, highlighting that reactive aggression is a response to some form of provocation and is usually accompanied by anger. In contrast, proactive aggression is usually not provoked and is more thoughtful than emotional.

The classification of *proactive* and *reactive* aggression is used interchangeably with the *instrumental – affective (hostile)* aggression classification (Anderson & Huesmann, 2003) and although the classification has been empirically supported (Dodge & Coie, 1987; Pulkkinen, 1996; Vitaro, Brendgen, & Tremblay, 2002), it has its limitations. In particular, distinguishing between reactive or proactive aggression is challenging given that it requires identifying whether or not provocation has occurred. Another difficulty relates to the fact that there are individual differences in what is perceived as provocation. Attributional or perceptual biases are of particular relevance. For example, research has shown that reactively aggressive children are more likely to perceive ambiguous situations as threatening or hostile (Dodge & Coie, 1987; Vitaro et al., 2002).

Another distinction is that of *impulsive* versus *premeditated* aggression. Impulsive aggression is conceptualized as being automatic and non-cognitive or thoughtless, and occurs without giving consideration to consequences. In contrast, premeditated aggression is deliberate and slow, less emotional, and more proactive rather than reactive.

As is apparent, the dimensions used to differentiate what have been considered by some to be different subtypes of aggression are not distinct. Thus, although some have considered the identification of subtypes of aggression to be useful (Gendreau & Archer, 2005), others such as Anderson and colleagues (e.g., Anderson, 2000; 2002; Anderson & Huesmann, 2003; Bushman & Anderson, 2001) have highlighted that the identified subtypes are not mutually exclusive. As with other antisocial behaviours, different types of aggressive acts tend to co-occur. Thus, considering the identified differentiating dimensions of aggressive behaviour to classify distinct subtypes can result in unnecessary conceptual and empirical confusion and problems. For example, Anderson and Huesmann (2003) have noted that instrumental aggression can also include hostile affect as is true for hostile aggression. It is also possible for some proactive aggression to be distinctly emotional. Moreover, frequent engagement in aggressive acts promotes automatic and non-conscious processing such that instrumental consideration of potential consequences can occur without awareness (Bargh & Pietromonaco, 1982; Schneider & Shiffrin, 1977).

On balance, given the inherent difficulties involved in sub-typing aggression, Anderson and Huesmann (2003) have concluded that it is more useful to characterize aggressive behaviour as varying along identified dimensions. Such a strategy acknowledges the multidimensional nature of aggressive behaviours. It is also consistent with current conceptualizations of animal cruelty (c.f., Kellert & Felthous, 1985). Identified dimensions of aggression include: (i) the degree of hostile affect versus agitated affect; (ii) the degree to which the primary or ultimate goal is to cause harm to the victim versus an instrumental goal of deriving a profit or reward; and (iii) the degree to which the likely consequences were considered, reflecting whether the aggressive behaviour was premeditated (thoughtful, deliberate, slow and instrumental) or impulsive (automatic, fast, affect-laden).

The argument to examine aggression according to quantitative differences on multiple dimensions rather than as categorical differences is consistent with the understanding that there are co-occurrences between different antisocial behaviours and that it is most useful to examine aggression within the larger framework (Dodge et al., 2006) as opposed to breaking it down into separate categories. It is also consistent with the increasing evidence that human aggression is best viewed as a heterogeneous category of human behaviour (Dodge et al., 2006), perhaps best simply defined as "acts intended to harm others" (p. 722) since "no single statement can adequately bound the acts that we would want to describe as aggressive behaviors" (p. 722). Instead, a judgement is necessary regarding aspects of the behaviour such as cues to intent, biological arousal, outcome potential, and the social context of the act(s).

Chapter summary

The adverse consequences of antisocial behaviours to individuals, communities, and society as a whole are highlighted in this chapter. One important research outcome is the consistent finding that antisocial behaviours are stable over time, from childhood through to adulthood. Increased crime rates and homicides over the past 40 years have been documented in industrialized societies. Of note, whilst chronically violent individuals constitute a relatively small group in society, they are responsible for more than half of all crimes. This finding has directed the research focus away from violent and antisocial *behaviours* toward a focus on the *individuals* who engage in those behaviours.

Research has also revealed that aggressive behaviours mostly occur within the context of other antisocial behaviours. Furthermore, individuals who have committed a variety of antisocial acts, and who are frequent offenders throughout their childhood and adolescent years, are more likely to progress to more serious violent acts.

The final part of this chapter reviews conceptualizations of the antisocial behaviour and aggression constructs. Aggression is best understood as being one behaviour that is subsumed within the heterogeneous antisocial construct most simply defined by Dodge and colleagues (2006) as "acts intended to harm others" (p. 722). However, as discussed, arriving at an agreed upon conceptualizations of these terms has not been an easy task. Whilst some authors have put forth subtypes of aggression such as *instrumental* versus *hostile* aggression, or *proactive* versus *reactive*, others have argued that such subtypes only add confusion to an already complex construct. Given that the subtypes are not necessarily mutually exclusive or distinct some have argued that it is more useful to characterize aggressive behaviours as varying along a number of identified dimensions including, for example, degree of hostile versus agitated affect or the degree to which likely consequences were or were not considered. Conceptualizing aggression according to quantitative differences on multiple dimensions is consistent with increasing evidence that human aggression is a heterogeneous category of human behaviour.

To better understand and work toward treatment and prevention of antisocial behaviour, the developmental trajectories of individuals identified as being antisocial has been of particular focus for researchers and professionals in the area, as will be reviewed in the next chapter.

4
The Development of Antisocial Behaviour

Prior to proceeding to a review of research examining the development of antisocial behaviour from infancy through to adulthood, this chapter provides an overview of the key issues and considerations relating to the development of aggression, violence and antisocial behaviour. The final sections of this chapter focus on the documented prevalence and stability or course of antisocial behaviours.

Key issues and considerations

On the basis of the large body of aggression research that has been conducted, there exists significant understanding regarding the development of aggressive and related behaviours from infancy through to adulthood. In their review, Loeber and Hay (1997) identified what they referred to as key issues in the development of aggression and violence from childhood to early adulthood. These include that: (i) The manifestations of aggression change dramatically in form and function throughout childhood, adolescence and early adulthood. These changes also differ depending on whether one is male or female. (ii) Whereas aggression is, to some degree, age-normative, particularly for boys, variations in normal development can identify highly aggressive individuals. (iii) Although most violent individuals have a developmental history characterized by an escalation in the severity of aggression, a minority of individuals experience a late onset of violence without such a history. (iv) Violence has both cognitive and emotional antecedents. Among the earliest manifestations of violence antecedents are difficult temperament and poor emotion regulation (see definitions of temperament and emotion regulation below). Relevant cognitive factors include: social cognitive deficits, mental scripts, attitudes that are favourable to antisocial behaviour, rejection sensitivity, and inflated self-esteem. These variables will be discussed in some detail in the chapters that follow.

Temperament refers to an internal disposition that influences relatively stable styles of behaving over time and across situations. For example, an inhibited temperamental style influences behaviour such as avoidance of unfamiliar people, objects, and situations. In contrast, an uninhibited style influences spontaneous approach behaviours toward novel persons, objects, and situations (Schwartz, Wright, Shin, Kagan, & Rauch, 2003).

Emotion regulation involves processes that enable us to be aware of our emotions as well as processes that enable us to monitor, evaluate and change our emotions in order to achieve our goals in a manner that is appropriate for the particular situation.

Other considerations include classification issues related to the broad range of behaviours considered under the rubric of antisocial behaviour. The Diagnostic and Statistical Manual for Mental Disorders – IV-TR (DSM-IV-*TR*) (2000) sub-types antisocial youth according to the pattern of their behaviours. Youth whose behaviour pattern can be described as argumentative, non-compliant, and oppositional are classified within the Oppositional Defiant Disorder category. In contrast, youth whose pattern of behaviour is predominantly characterized by aggressive, destructive, deceitful, and norm-violating behaviour are classified into the Conduct Disorder category. Other distinctions have been drawn between what have come to be referred to as *reactive aggression* versus *proactive aggression*. As noted in the previous chapter, the former classifies behaviours that occur in response to provocation that is real or perceived, and the latter classifies behaviours that are premeditated or that have the goal of instrumental gain. Another issue concerns the relationship between antisocial behaviour and Antisocial Personality Disorder. Some consider chronic patterns of severe antisocial behaviour and antisocial personality disorder to be indistinguishable. Personality trait profiles and age of onset of an antisocial behaviour pattern are the basis for other taxonomies (Frick & Viding, 2009). These issues will be discussed further in this chapter beginning with a review of the development of antisocial behaviour from infancy through to adulthood.

Development

Infancy and toddlerhood

According to Loeber and Hay (1997), the earliest manifestations of aggression occur in the infant's early interactions with their social world. Most infants show signs of frustration and rage, and there appears to be no sex difference at this early age. Research has indicated that it may not be until at least four-months of age that facial displays of anger begin to take on an adaptive function in the way of social signals (Sternberg & Campos, 1990). Whether aggression has developed by this early age or not is particularly difficult to determine, given the challenges of determining the presence of intent to harm at such an early age. In terms of anger expression, by the first year of life, stability in individual differences begins to become apparent (Stifter, Spinrad, & Braungart-Rieker, 1999), and observations of peer-directed aggression have been reported (Hay, Nash, & Pederson, 1983). Examples include protests and aggressive retaliations to provocations, such as an infant having their toys taken by another (Anderson & Huesmann, 2003). By the ages of two to four years, verbal aggression increases whilst physical aggression decreases (Tremblay, 2000).

Regarding antecedents of aggression or triggering circumstances, in infancy most involve physical discomfort or the need for attention, but by the fourth and fifth years, peer conflicts and conflicts over material possessions become more commonplace (Fabes & Eisenberg, 1992). Indeed, even as early as 12 to 18 months of age, as many as half of all peer exchanges have been shown to involve conflict which commonly occurs to obtain instrumental goals (Anderson & Huesmann, 2003).

During the developmental periods of infancy and toddlerhood, few sex differences have been documented, however, infant girls have been found to be better able to regulate their emotional states, and boys have been reported to express both their positive and negative emotions at higher rates (Weinberg, Tronick, Cohn, & Olson, 1999). Sex differences have been reported to be pronounced during the preschool years, particularly with regard to physical aggression (Loeber & Hay, 1997; Underwood, 2003).

Childhood

The overall pattern of aggression expression indicates that from early to middle childhood for both boys and girls, there is a decrease in aggression along with an increase in interpersonal skills. These changes are understood to result from children being socialized out of behaving aggressively, and they describe a normative trend (Anderson & Huesmann, 2003). During these years, sex differences in aggression become more prominent, particularly during the third to sixth years. Indirect aggression increases for girls and physical aggression increases for boys. Both males and females are equally likely to become more verbally aggressive (Crick & Grotpeter, 1995; Laperspetz & Bjorkqvist, 1992).

In addition to reflecting increasing socialization to societal norms, the overall decrease in aggressive behaviours documented during this developmental period involves the development of self-control over emotion otherwise referred to as emotion regulation (Keenan & Shaw, 2003). The development of the ability to delay gratification, including development of the cognitive strategies of distraction and the anticipation, or mental representation of delayed rewards, contributes to the decrease in aggression (Mischel, 1974). Other developing abilities include those of effortful control (Eisenberg, Champion, & Ma, 2004), perspective-taking (Selman, 1980), empathy (Zahn-Waxler, Radke-Yarrow, & King, 1979), and emotion processing (Schultz, Izard, & Bear, 2004). Of particular relevance, it has been found that attention-shifting strategies including, for example, ignoring frustrating stimuli, are associated with the ability to control anger at 42 months and with teacher-reported externalizing problems at age six (Gilliom, Shaw, Beck, Schonberg, & Lukon, 2002). According to Eisenberg and Fabes (1999), during these elementary-school years, emotion regulation develops from externally controlled regulation to internally controlled cognitive mechanisms or strategies. This transition to cognitive regulation is considered to partly explain decreases in aggressive behaviours with maturation. Another factor considered to contribute to decreases in aggression or antisocial behaviours is peer feedback, which works toward extinction of such behaviours (Dodge et al., 2006).

With development, there are also changes in the functions served by aggressive or antisocial acts. In particular, aggression becomes more likely to be directed at specific relationships (Dodge et al. 2006), and its form becomes more hostile, compared to the relatively non-social, and more instrumental aggression more typical of the preschool period. Covert forms of antisocial behaviour including lying, stealing, and cheating become more frequent during the elementary-school years (Loeber, Farrington, Stouthamer-Loeber, & van Kammen, 1998). During these years, characteristic elicitors of antisocial behaviour include threats and insults. Also, there is the emergence of intentional aggressive or antisocial motives, which relates to an increase in angry responding or retaliatory behaviours (Gifford-Smith, & Rabiner, 2004; Schwartz, McFayden-Ketchum, Dodge, Pettit, & Bates, 1998).

Child-onset antisocial behaviour

Of note, for a subset of children otherwise referred to as the *childhood-onset group* or the *early starters*, the developmental trend is in the opposite direction to that described above, with their most dangerous period being late adolescence and early adulthood. For these children, mild conduct problems begin as early as the preschool or early elementary-school years.

Indeed, the DSM-IV-*TR* (2000) distinguishes between childhood and adolescent-onset Conduct Disorder. The childhood-onset subtype is defined by the onset of at least one criterion characteristic of Conduct Disorder prior to age 10. Children presenting with behaviours including animal cruelty, firesetting, fighting, weapon use, and vandalism during this period usually meet the criteria for diagnosis of Conduct Disorder. Reflecting the empirical literature, the Diagnostic and Statistical Manual of Mental Disorders describes childhood-onset individuals as being more likely to have persistent Conduct Disorder and to develop adult Antisocial Personality Disorder than are those with the adolescent-onset subtype (Loeber & Hay, 1997). In contrast, the adolescent-onset subtype is defined by the absence of any criteria characteristic of Conduct Disorder prior to age 10. Individuals classified as meeting the criteria for diagnosis into this category are described as less likely to display aggressive behaviours. They are also less likely to have persistent Conduct Disorder or to develop adult Antisocial Personality Disorder.

Research has indicated that the severity and rate of aggressive behaviours increases throughout childhood and adolescence for children classified within the childhood-onset category (Lahey & Loeber. 1994). Their behaviours include violent and criminal behaviours such as forced sex, breaking and entering, and mugging (Conduct Problems Prevention Research Group, 1992; Moffitt, 1990; 1993; Patterson, Reid, & Dishion, 1992. In addition to behaving more aggressively, these children have been reported to be more likely to carry weapons. Not surprisingly, the consequences of their behaviour are more severe and they are more likely to continue to show antisocial and criminal behaviour in their adult years (Anderson & Huesmann, 2003; Frick & Viding, 2009).

According to Lahey and Loeber (1994), these children develop antisocial behaviours to the extent that they are perceived as problematic by parents, teachers, and peers. This is particularly true for boys. A developmental path can be identified that is characterized by conduct problems and begins with oppositional behaviours such as temper tantrums, irritability, and argumentative behaviours. While such behaviours are somewhat normative during early childhood, if they occur around age eight or beyond the childhood years, they are considered markers of a problematic pathway (Achenbach & Edelbroch, 1983; Loeber, Lahey, & Thomas, 1991).

Adolescence

Although most longitudinal studies show an overall decrease in antisocial behaviour as children transition into adolescence, the adolescent period is marked by an increase in serious acts of violence. This increase has been explained by a number of factors, including the broadening of antisocial behaviours to new contexts (Dodge et al., 2006). During this time, major changes occur in the level and pattern of antisocial behaviour such that more intense aggressive acts, including acts likely to cause serious injury and even death, become more frequent. Another change is that during adolescence, compared to earlier periods, peer groups in schools are more likely to engage in collective violence. During these years, peer groups tend to comprise individuals of similar levels of aggression and antisocial tendencies. Also, during adolescence, more organized gangs emerge. Youths who join gangs are more likely to become involved in violent crime. Moreover, gangs in schools are often associated with a gang presence in the community. Moreover, as dating begins, violence between males and females increases in prevalence. Regarding sex differences, whereas adolescence is more likely to mark a decrease in violence for females, the opposite is true for boys (Loeber & Hay, 1997).

In reviewing available trends for violent behaviour, Dodge and colleagues (2006) note that the risk of serious violent offences, including robbery, rape, and aggravated assault, rise sharply from ages 12 to 20. More than half of the individuals who become involved in serious violent offending before the age of 27 begin offending as early as 14, and the vast majority have committed their first offence by the age of 21.

The overall picture is one in which the developmental pathway for serious violence begins with minor forms of delinquent behaviour that tends to occur during the middle childhood years. For a small proportion of the population, these behaviours progress to more frequent and more violent offending by the age of 17. In contrast, the vast majority of offenders desist from crime in early adulthood (Elliot, 1994).

Adolescent-onset antisocial behaviour

With regard to the child versus adolescent-onset antisocial behaviour pattern, in contrast to individuals in the former group, those who exhibit later-onset patterns of antisocial behaviour (otherwise referred to as the *adolescent-onset* pattern) do not show significant behaviour problems in childhood. Rather, their delinquent and antisocial behaviour begins with the onset of adolescence (Moffitt, 2003). Also compared to children whose antisocial behaviour began earlier, the adolescent-onset group is less likely to continue their antisocial behaviour pattern into adulthood. Thus, according to Frick and Viding (2009), individuals in the childhood and adolescent-onset subgroups of antisocial youth show very different trajectories of antisocial behaviour in terms of both the onset patterns and the life-course of their antisocial behaviour. The behaviours of the adolescent-onset group have been described as an exaggerated form of adolescent rebellion (Moffitt, 2003). Those in this onset group are understood to behave in an antisocial way in response to antisocial peer group pressure in an attempt to gain a sense of maturity. Their behaviour is not likely to persist beyond adolescence. Nevertheless, their adolescent behaviour may incur adverse consequences for their adult years, including, for example, a criminal record and dropping out of school. (Frick & Viding, 2009).

Child versus adolescent-onset of antisocial behaviour

Reflecting their different trajectories, the child-onset group is also referred to as the *life-course persistent group*, and the adolescent-onset group has been referred to as the *adolescence-limited group* (Moffitt, 2003; 2006). Of note, dispositional and contextual risk factors differ depending upon whether problem behaviours have early (childhood) or late (adolescence) onset. Such factors include lower intelligence, attention deficits, emotion regulation difficulties, and higher impulsivity for the childhood-onset group. Homes with greater family instability, more conflict, and problematic parenting strategies are also more strongly predictive of childhood-onset antisocial behaviour, compared to the adolescent-onset group. Other factors characteristic of child-onset include the child being born to a single mother, marital changes in the family, marital conflict, physical punishment, and a low maternal level of education (e.g., Bor, McGee & Fagan, 2004). Thus, as summarized by Frick and Viding (2009), the development of the problem behaviours characteristic of those in the child-onset group involves interactions between problem child behaviours (e.g., impulsive behaviours) and inadequate child rearing (e.g., poor supervision). The child consequently develops poor social relations with family members as well as with people outside the family (e.g., peers). This maladjustment is likely to spill over into other developmental periods, leading to enduring vulnerability.

This early-onset pathway contrasts with adolescent-onset group for whom problems are more likely to be limited to the adolescent developmental period. The risks for the adolescent-onset group are fewer and include individual attitudes supportive of problematic behaviours; low levels of emotional, social, and educational competencies; parental and family attitudes supportive of antisocial or criminal behaviours; low levels of family, school, and community connectedness; engagement with delinquent peers or gangs; and peer conflict.

At a community level, risks for late-onset antisocial behaviour include living in a low-socioeconomic-status area, low levels of neighbourhood attachment, exposure to crime and drugs, and exposure to high social disorganization (including behaviours such as showing little regard for other people's property), and high mobility (e.g., Green, Gestern, Greenwald, & Salcedo, 2008; Herrenkhohl, Maguin, Hill, Abbott, & Catalano, 2000). Generally, there is a more positive developmental trajectory for adolescent-onset antisocial behaviour.

In the developmental psychopathology approach that they have taken in their review of antisocial behaviour, Frick and Viding (2009) have highlighted the clearly identifiable subgroups of antisocial youth. The differences in course and correlates are particularly marked between the child-onset and adolescent-onset groups. Specifically, adolescent-onset antisocial behaviour can best be characterized as an exaggerated form of normative adolescent development (Moffitt, 2003). In contrast, the behaviour demonstrated by the child-onset group is characterized by significant adjustment problems across multiple stages of develop-ment. For this group, there are also more risk factors of a more severe nature.

The child-onset group has been further divided into youth with and without Callous-Unemotional traits (see Chapter 6 for a detailed discus-sion of these traits). The sub-group with Callous-Unemotional traits shows more severe conduct problems and a temperamental style char-acterized by preference for dangerous and novel stimuli, also referred to as *low behavioural inhibition* (Kagan & Snidman, 1991). Such youth also show a response style that is reward-oriented and lacking reactivity to signs of distress in others. Interestingly, low behavioural inhibi-tion has been associated with compromised conscience development (Kochanska, 1993). Children with this temperamental style are less likely to experience empathic arousal associated with distress in others (Blair, 1999). Thus, deficits in emotional reactivity of the sort demon-strated by youth with an uninhibited behavioural style compromise their ability to develop and experience normative and situationally appropriate guilt, empathy, and conscience. Frick and Viding (2009) have argued that such deficits, when extreme, are likely to be predictive of Callous-Unemotional traits and Antisocial behaviour.

In contrast, youth with child-onset Antisocial behaviour who do not display Callous-Unemotional traits show a markedly different aetiology. Dispositional factors include temperamental impulsivity as well as anxiety, poor emotion regulation, and low verbal intelligence. Family dysfunction is the most characteristic contextual factor. Moreover, the deficits in empathy and guilt characteristic of the Callous-Unemotional group are absent. The strong association between dysfunctional parenting experiences and antisocial behaviour suggests that the predominant pathway of development of Antisocial behaviour for this group is inadequate socialization and emotion regulation competency development. Indeed, in contrast to the callous-unemotional youth, these youth demonstrate strong emotional reactivity and a deficit in the competencies needed for regulating this reactivity. Impulsive tendencies, along with problematic emotion regulation, likely largely explain the Antisocial behaviour of youth without Callous-Unemotional traits (Frick & Viding, 2009).

Adulthood

Normatively, the increase in aggression during the adolescent period is followed by a decrease in early adulthood. Between the ages of 18 and 25, self-report data indicate a near absolute absence of new cases of Antisocial behaviour (Dodge et al., 2006). Through their examination of adult trajectories of a sample of delinquents, Sampson and Laub (2003) reported a further decline in crime after the age of 35 in groups of early offenders. Farrington (1991) reported a similar trend based on U.K. self-report data. Official records have confirmed these trends (Blumstein, 2000). Evidence supports the conclusion that the decline in violence and aggression in adulthood is predicted by the establishment of stable family life and work or career (Rutter, 1989). For example, in longitudinal research, Glueck (1959) found that strong ties to adult institutions, including work and family, predicted desistance from subsequent crimes among those with a history of delinquency. Marital cohesiveness was also found to play a protective role (Sampson & Laub, 1990). In addition to a decrease in aggression, early adulthood is characterized by the development of sex differences in physically violent behaviour. These differences are reflected in different homicide rates between males and females (Anderson & Huesmann, 2003).

Indications that there is a decrease in antisocial behaviour during adulthood, although positive, have also been argued to be problematic given that the trends and prevalence rates on which these conclusions are based exclude information regarding child or spousal abuse. Given the high prevalence rates of such abuse, the overall documented prevalence of aggression and violence in adulthood is likely to be underestimated. One study conducted by Huesmann, Eron, Lefkowitz & Walder (1984) reported continuity between peer-directed aggression at age eight, and spousal as well as child abuse in adulthood at age 30, suggesting that these forms of violence have antecedents similar to those for other forms of violence.

Of significant concern is the high prevalence of disorders of aggression and antisocial behaviour which are more prevalent amongst males and are relatively stable across the early adult years until the age of 40 when the prevalence decreases somewhat. There is also variation in prevalence throughout the earlier developmental periods.

Prevalence

In general, studies report that the prevalence of physical aggression is highest during the younger years and decreases during adolescence. For example, in some of his work, Loeber found that physical fighting decreased in prevalence from early adolescence onward, with an accelerated rate of decrease between ages 14 to 16. In contrast, however, more serious violence has shown the opposite trend, increasing with age, particularly during the adolescent years. According to Loeber and Hay (1997), the prevalence of violence tends to follow the "well-known age-crime curve, peaking in late adolescence or early adulthood and then decreases with age" (p. 381). Based on criminal records, the peak arrest rates for criminal violence vary little by sex (Loeber & Hay, 1997).

The prevalence of Conduct Disorder is such that it is among the most common reasons for referral to mental health services (Robins, 1991). Estimates of the prevalence of Conduct Disorder among children in the 1970s ranged from approximately 4% to 10% (Rutter, Cox, Tupling, Berger, & Yule, 1975; Rutter, Tizard, & Whitmore, 1970). Rates appear to have increased in recent years but vary widely depending on the population sampled and methods of ascertainment. The overall prevalence of clinical diagnoses has been reported to range between 2% and 6%. The majority of youth diagnosed with Conduct Disorder meet the diagnostic criteria for Antisocial Personality Disorder during their adult years (Gelhorn et al., 2007). The DSM-IV TR features for Antisocial Personality Disorder include a pervasive pattern of disregard for the rights of others that begins prior to 15 years of age.

Individuals who meet the Diagnostic and Statistical Manual of Mental Disorders criteria for the disorder show a failure to conform to social norms. They demonstrate regular violation of the law with behaviours including the wilful destruction of property, stealing, and illegal forms of employment. Behaviours particularly characteristic of Antisocial Personality Disorder include deceit and manipulation, as indicated by frequent lying to others or the conning of others for personal pleasure or gain. Also characteristic are impulsivity and failure to plan ahead, irritability, and aggression marked by repeated physical fights or assaults, a disregard for one's own safety and for that of others evidenced as reckless behaviour, and consistent irresponsibility as indicated by repeated failure to retain paid employment or to honour financial obligations. Adults diagnosed with Antisocial Personality Disorder also show a lack of remorse most clearly indicated by their indifference to, or rationalization of, their mistreatment of others (American Psychiatric Association, 2000). The clinical population prevalence of Antisocial Personality Disorder has been reported to be 3.6% (Lenzenweger, Lane, Loranger, & Kessler, 2007) and around 1% in community samples (Torgersen, Kringlen, & Cramer, 2001). A male preponderance has been documented for clinical samples and no sex difference in community samples. Regarding stability, the behaviours characteristic of Antisocial Personality Disorder have been found to dissipate somewhat around the age of 40 (Hare, McPherson, & Forth, 1988).

Whilst stability has been noted as most characteristic of pathological levels of aggression and antisocial behaviour, age of onset has been documented to be an important marker of stability as has the degree of aggression ranging from low to high aggression. This literature will be summarized below.

Stability

Aggression in the early years has been found to be a risk marker for later violence. It is also evident that individual differences in the propensity to behave aggressively become apparent by as early as the preschool and elementary-school years (Huesmann, et al., 1984; Loeber & Hay, 1997). Long-term stability has been demonstrated in many longitudinal studies (e.g., Farrington 1991; 1994; Haapasalo & Tremblay, 1994; Tremblay et al., 1991). Research has shown that childhood antisocial behaviour is predictive of adult antisocial behaviour with as many as 75% of youth diagnosed with Conduct Disorder progressing to Antisocial Personality Disorder in adulthood (Gelhorn et al., 2007).

Thus, according to longitudinal research, antisocial behaviour in adulthood begins in childhood (Gelhorn et al., 2007; Loeber, 1982). Indeed, between 14% and 54% of children who show severe conduct problems, but who do not meet the diagnostic criteria for Conduct Disorder, will meet the criteria for Antisocial Personality Disorder in adulthood (e.g., Lahey, Loeber, Burke, & Applegate, 2005; Moffitt & Caspi, 2001). The stability of antisocial behaviour over time is so strong that it has been compared to the stability shown over time for intelligence, a construct known to have high stability (Lyons-Ruth, 1996).

Of note, high and long-term stability has been demonstrated across the entire severity range of aggression (Huesmann & Moise, 1998) but seems to be highest at the extremes of the aggression distribution with those who display the lowest levels and those who display the highest levels demonstrating the highest stability. Most violence seems to "erupt in youths who have been aggressive earlier in life" (Loeber & Hay, 1997, p. 384).

When tracing back over the earlier behaviour of known violent adults, the findings have revealed that violent offenders were highly aggressive during their earlier years. Farrington (1978) showed that seven out of ten men who were charged with violent crime by the time they were aged 21 had been rated as highly aggressive during their early to middle adolescent years. Farrington (1995) showed that almost 75% of youths convicted of juvenile crimes when aged between ten and 16 years were convicted again between the ages of 17 and 24 years, and as many as 50% of those convicted for juvenile crimes were convicted again when between ages 25 and 32. Given the comorbidity across different antisocial behaviours, it is not surprising that longitudinal research has also shown that aggression in early adolescence predicts other forms of aggression later in life, including spousal or partner abuse as well as convictions for violent offences (Loeber & Hay, 1997).

Demonstrating stability as well as sex differences, Huesmann, et al.'s (1984) 22-year follow-up study involved a group of men and women who had been rated as aggressive by their peers during their late childhood years. The findings revealed that the men were more likely to commit serious criminal acts, to abuse their spouses, and to drive while intoxicated, and the women were more likely to punish their children severely, compared to men and women who were not rated as aggressive as children. These same researchers demonstrated high stability in aggression across generations.

Such data indicate that there is a high degree of stability in individual differences from adolescence through to the adult years. The continuity of aggression into adulthood manifests as varying forms of criminal behaviour, including drunk driving and moving traffic violations. Pulkkinen (1990) has referred to such manifestations as a "relapsed lifestyle".

Chapter summary

Following an examination of the key issues related to the development, prevalence, and course of antisocial behaviour, the chapter reviews the research examining the development of antisocial behaviour as classified into the main developmental areas of infancy and toddlerhood, childhood, adolescence, and adulthood. Issues considered include the different manifestations of aggression, depending on developmental period and sex, differences in severity of aggression, antisocial behaviour, and violence, depending upon whether there is early childhood onset or later onset.

During the infancy and toddlerhood developmental periods, the emotions of anger and frustration are first expressed. By the fifth year of life, the earliest signs of aggression become evident. During the childhood years, the prevalence of aggression decreases as interpersonal and emotion regulation skills develop. Sex differences become quite prominent during these years. By the adolescent developmental period, there is an overall decrease in antisocial behaviour. However, the research has shown an increase in the prevalence of serious acts of violence during this period. This can be explained in part by the increased significance and influence of peer groups and gangs during the adolescent developmental period. There are opposing sex trends during this period with an increase in violence for boys and a decrease for girls. During adulthood, aggression levels decrease, particularly around the age of 35. Stable career and family life are credited as playing an important role in this decrease.

One of the research findings that has been most valuable in understanding the development of antisocial behaviour is the very different trajectories of individuals depending upon whether their antisocial behaviour begins during childhood (the early starters) or during adolescence (the late starters). Reflecting the difference in trajectory are the labels given to these two groups, as the *life-course persistent group* or the *adolescent limited group*, respectively.

The chapter ends with discussion of the prevalence and stability of antisocial behaviour with particular reference to the disorders of antisocial behaviour, including Conduct Disorder and Antisocial Personality

Disorder. These two disorders are among the most prevalent disorders classified in the *Diagnostic and Statistical Manual of Mental Disorders*, and they are also both more prevalent among males.

The stability of antisocial behaviour over the course of development is extremely high, with as many as 75% of youth diagnosed with childhood Conduct Disorder progressing to Antisocial Personality Disorder in adulthood, and between 14% and 54% of children who have shown severe conduct problems progressing to Antisocial Personality Disorder in adulthood. Such knowledge reinforces the importance of treating antisocial behaviours during the childhood years in an effort to prevent them from progressing into more severe forms of adult antisocial behaviour.

As has been stated a number of times, antisocial behaviour is generally agreed upon to be a heterogeneous construct. Along with incorporating aggression and its developmental trajectories, it incorporates animal cruelty and its developmental trajectories. Theoretical understandings of aggression and animal cruelty and their development are the subject of the next chapter.

5
Theoretical Accounts of Aggressive Behaviour and Animal Cruelty

Given the longer history and larger literature focussing specifically on aggressive behaviours as compared to individuals who behave aggressively, it is not surprising that major theories in this area relate specifically to aggressive behaviour. However, the change in focus from understanding aggressive behaviours to better understanding the developmental trajectories of antisocial individuals has resulted in a corresponding change in theoretical focus. This chapter provides an overview of notable theoretical understandings of aggression and animal cruelty. The more recent focus on aetiological and developmental factors regarding the developmental trajectories of antisocial individuals will be the focus of subsequent chapters.

Theoretical accounts of aggressive behaviour

There are several theoretical understanding related to antisocial behaviour (Anderson & Huesmann, 2003). These include the emergence of the cognitive neo-association theory put forth by Berkowitz (1989, 1993), who emphasized enduring associations between affect, cognition, and situational cues as well as their importance for understanding antisocial behaviour. Another notable advance was contributed by social cognitive models (Bandura, 1973; 1983; Mischel, 1973). Anderson and Bushman (2002) have integrated the constructs forming much of aggression research into the General Aggression Model. This model incorporates research looking at schemas and scripts, interpretation, decision making, and action (e.g., Anderson & Bushman, 2002; DeWall & Anderson, 2011). According to the model, cognitions provide the foundations for stability in behavioural tendencies seen across situations. Cognitions also provide the foundations for situational specificity of behaviour. These models will be discussed in more detail below.

Cognitive neo-association theory

This theory developed out of the frustration-aggression hypothesis put forth by Dollard and colleagues in 1939. It was an hypothesis that constituted one of the first major theoretical perspectives specific to aggression, and it dominated debates about aggression throughout the 1980s. It posited a drive theory and argued that aggression is an inevitable response to perceived goal blocking. It was proposed that frustration results in aggression and that aggression is inevitably preceded by frustration. Berkowitz (1962) called this absolute position into question with empirical work showing that frustration does not always lead to aggression. However, frustration does create a readiness to aggress through the arousal of anger (Berkowitz, 1993).

According to the cognitive neo-association theory, aversive events such as frustrations, provocations, unpleasant odours, uncomfortable temperatures, and loud noises produce negative affect. The negative affect, in turn, stimulates associated thoughts, memories, motor reactions, and physiological responses which result in a flight response if fear is aroused, or a fight response if anger is aroused. The theory postulates that the negative affect that results from aversive stimuli is initially undifferentiated. Over time, along with situational cues, the negative affect results in the development of a network of cognitive structures. Once developed, these structures influence the ways in which individuals perceive and interpret negative affect, which consequently influence behaviour. In addition to incorporating the frustration-aggression hypothesis, the neo-association theory postulates that negative affect constitutes a link between aversive events and aggressive tendencies (Anderson & Bushman, 2002).

The neo-association theory, however, does not incorporate social processes which Albert Bandura argued are significant for humans. His model therefore places such processes at the centre of his theory on aggression.

Social cognitive models

Bandura (1978) argued that aggression is a multi-faceted phenomenon that serves potentially diverse purposes. It can also have many determinants. Thus, to be a complete theory, a theory of aggression must be able to explain how aggressive patterns develop, what provokes aggressive behaviours, and what sustains such behaviours once they have been initiated. Bandura acknowledged that biological and genetic factors play an important role but argued that their role in humans is significantly less important compared to other species. To this end, Bandura (1973) hypothesized that social learning processes are integral to the acquisition of aggressive behaviours.

Social learning processes including observational learning and cognitive processes, such as self-judgements and moral disengagement, have been shown in research to be useful in understanding individual differences in aggressive behaviour. Of note, Bandura (1978) identified three major pathways through which social learning processes take place in modern societies. Such pathways include the imitation of aggression modelled and reinforced by family members. A second pathway is from sub-cultures in which aggression is regarded as a valuable attribute (such as gangs and aggressive peer groups), and a third pathway is from the symbolic modelling that comes via the mass media. For this pathway, Bandura (1978) has noted the considerable evidence supporting the effect of televised violence on viewers. The effects include: (i) the teaching of aggressive styles of behaving, (ii) the altering of restraints over aggressive behaviour, (iii) the desensitization of, and habituation to, aggression and violence, and (iv) shaping people's conceptions of reality toward a more antisocial or aggressive behaviour baseline.

In addition to the extensive writings and empirical work by Bandura and colleagues (e.g., Bandura, Barbaranelli, Caprara, & Pastorelli, 1996; Bandura, Underwood, & Fromson, 1975; Bandura & Walters, 1959), the work by Patterson and colleagues (1989) on understanding how family interactions influence the acquisition of aggressive behaviour is predominantly based upon processes put forth within the social learning perspective. Patterson's work will be discussed in more detail in Chapter 7.

In contrast to the neo-association and social cognitive models, the General Aggression Model put forth by Anderson and Bushman (2002) places cognitive constructs at the centre, as will be detailed below.

General Aggression Model (GAM)

Anderson and Bushman (2002) describe the GAM as their attempt to integrate existing "mini-theories" of aggression. Key aspects of the GAM include proposals heavily based upon cognitive constructs. Accordingly, it is proposed that knowledge structures develop out of experience. Once developed, these structures influence perception at multiple levels ranging from the most basic level of visual pattern perception to the more complex level of behavioural sequences. In addition to such broadly ranging effects, research has shown that knowledge structures can become automatized with use and thus can operate unconsciously.

They can also be linked to affective states and beliefs as well as guiding individuals' interpretations of their environments and their behavioural responses in those environments. The GAM adopts the framework of the "person in the situation" (Anderson & Bushman, 2002). As such, the model focuses upon three main areas: (i) input provided by person and situational factors, (ii) cognitive, affective, and arousal routes through which the input variables have their impact, and (iii) the outcomes associated with appraisal and decision processes.

As with the separateness of the evolved literatures on human aggression and antisocial behaviour and that relating to animal cruelty, separate theoretical perspectives have been put forth. However, despite the apparent differences, it is not surprising that there is quite a bit of overlap between theories that attempt to explain animal cruelty and, in particular, the new wave of understanding relating to human aggression that attempts to shed light on the characteristics of the individuals rather than their behaviour.

Theoretical accounts of animal cruelty

The two major theoretical perspectives that have guided animal cruelty research are the Violence Graduation Hypothesis and the Deviance Generalization Hypothesis. According to the former, animal cruelty in childhood graduates to violence toward humans in adulthood. That is, the presence of animal cruelty at one developmental period is predictive of interpersonal violence at a later period (Ascione & Lockwood, 2001). In contrast, according to the Deviance Generalization Hypothesis, animal cruelty is one of many forms of antisocial behaviour that can be expected to arise from childhood onward. According to this theory, widely ranging deviant behaviours, including criminal behaviours, are positively correlated with one another, either because one form of deviant behaviour leads to involvement in another or because certain different forms of deviance have the same underlying causes (Arluke et al., 1999; Ascione, 1999). This line of thinking is consistent with current understandings of antisocial behaviour, given the findings that (i) the same antisocial individuals often engage in a variety of antisocial behaviours, and (ii) there is increased heterogeneity at the more severe end of the antisocial spectrum.

Antisocial behaviour risk factors

A large amount of research has been conducted within the framework of the major theoretical approaches outlined above. This research has significantly facilitated understanding of the aetiology of aggressive and antisocial behaviours at various levels. As with all complex behaviours, causal factors are many for antisocial behaviours and range from biological and genetic factors to environmental and contextual factors. More recently, the aim of research in the area has been to better understand the complex interactions among factors previously recognized as important aetiological factors.

Existing research includes determination of risk factors for the development of antisocial behaviours, or factors that identify with some degree of accuracy the characteristics of individuals who engage in such behaviours. These include factors that are specific to the individual, such as biological predispositions, personality traits, and cognitive structures, such as schemas and scripts.

At a broader level, factors include the individual's social environment, such as peers, school, and the general community or neighbourhood. Other environmental or contextual variables, such as socio-economic status, exposure to violence, and situational variables, have also been identified as important (Reebye, 2005). Within a developmental framework, others (e.g., Greenberg, Speltz, & DeKlyen, 1993) have identified the importance of contextual factors best described as micro-level environmental factors, such as family stressors, parenting, discipline, and attachment as well as child dispositional factors, including temperament.

Other identified risk factors include marital conflict, single parenthood, low education, low income, and overcrowding (Lyons-Ruth, 1996). Risk factors have proven particularly useful for identifying subgroups of antisocial individuals. As noted in Chapter 4 on the development of antisocial behaviour, one such classification depends upon whether problem behaviours have early (childhood) or late (adolescence) onset. Another depends upon specific individual difference traits and characteristics including psychopathy and Callous-Unemotional traits. The chapters that follow review, in detail, research related to the above factors. Chapter 9 focuses on theoretical accounts and related empirical investigation of animal cruelty.

Chapter summary

This chapter provides an overview of theoretical understanding and focus in the areas of antisocial behaviour generally and aggression and animal cruelty more specifically. Whist earlier theories focussed on understanding aggressive behaviours, more recent research has examined the factors that characterize antisocial individuals. As with the more recent theoretical focus of human antisocial behaviour and aggression research, animal cruelty research has been guided predominantly by theories which attempt to shed light on the development and characteristics of individuals who are cruel to animals.

Theoretical proposals regarding the aetiology of antisocial behaviour and the development of antisocial individuals incorporate a range of risk factors including biological factors, emotional and cognitive processes, individual difference variables, and environmental variables. Although not explicitly noted in the literature reviewed in the subsequent three chapters, given the valid incorporation of animal cruelty behaviours and the individuals who engage in them, into the construct of antisocial behaviour, much of this research can be validly applied to the development of individuals who are cruel to animals. The relevance of much of this research to the development of individuals who are cruel to animals is further highlighted in Chapter 9, which reviews research that has explicitly focused on understanding these individuals.

6
Biological and Individual Difference Risk Factors

This chapter will review risk factors for the development of antisocial behaviour that can be classified as having a biological or heritable basis. This includes having a family history of Antisocial Personality Disorder, or whether one is born a female or male. It is relevant to note here the differences between *sex* and *gender* (see the insert below). This chapter will also review individual difference risk factors which include factors such as temperament and personality. Personality traits and temperamental styles can be conceptualized as internal dispositions that influence relatively stable styles of behaving over time and across situations. The chapter concludes with a review of an individual difference variable, otherwise referred to as "Psychopathy", that is more descriptive of abnormal or disordered functioning, as compared to the normative personality dimensions summarized in Table 1.

SEX refers to biologically-based differences.

GENDER refers to socially influenced characteristics and involves processes including gender stereotypes (widely held beliefs about characteristics deemed appropriate for males and females), gender roles (the reflection of gender stereotypes in everyday behaviour) and gender-role identity (the perception an individual holds about their self as relatively masculine or feminine in characteristics, behaviours and abilities).

Biological risk factors

Research examining genetic influences has been particularly comprehensive and has involved samples spanning the lifespan from early childhood to late adulthood (Dodge et al., 2006). This research has generally found that antisocial behaviour has moderate heritability particularly for life-course persistent antisocial behaviour and for individuals growing up in what have become identified as risky environments (discussed in more detail below) (e.g., Repetti et al., 2002). Other important constitutional variables include sex, age, and baseline arousal levels.

Sex differences

On the whole, there is substantial research indicating marked sex differences in antisocial behaviour. Boys outnumber girls on delinquency and externalizing behaviours. Moreover, these sex differences have been found across as many as 12 different cultures (Crijnen, Achenbach, & Verhulst, 1997) and from ages 5 through 21 (Moffitt, Caspi, Rutter, & Silva, 2001). Specifically relating to aggressive tendencies, males outnumber females by a ratio of about 10 to 1, and sex differences are noticeable as early as the preschool years (Loeber & Hay, 1997). This is particularly true for physical aggression, with differences emerging as early as three years of age and remaining stable into adolescence (Crick, Casas, & Mosher, 1997; Kingston & Prior, 1995) and even early adulthood (Stanger, Achenbach, & Verhulst, 1997).

A relatively large body of research has examined the prevalence of Conduct Disorder and its variance across cultures. Not surprisingly, such research has yielded similar findings to that relating to other antisocial behaviours. Generally, there is an earlier age of onset for boys, and the prevalence rate is three to four times higher compared to girls. Specifically, for males under the age of 18 years, rates range from 6% to 16%; for females, rates range from 2% to 9% (American Psychiatric Association, 1994). Rates of conduct problem behaviours for boys remain higher across the lifespan.

Sex differences have been reported to be more marked for more severe conduct problems that have an earlier onset. Longitudinal research (Moffitt & Caspi, 2001) examining a sample of 5- to 18-year-olds classified into groups of life-course persistent offenders and adolescence limited offenders found that the life-course persistent offenders group was heavily dominated by males with a male to female ratio of 10 to 1. The adolescent limited group was also more heavily weighted by males but the difference ratio was not as great with 26% of males compared with 18% of females.

Looking at more extreme antisocial behaviours, including homicide and aggravated assault, Anderson and Bushman (2002) have referred to United States Federal Bureau of Investigation (FBI) data for the years spanning 1951–1999. These data indicate that there is a ratio of male to female murderers of 10 to 1. Similar sex effects for aggression have been shown in laboratory-based studies. These sex differences are at least in part attributed to socialization effects (i.e., gender effects). However, there is also evidence to suggest that sex differences are, in part, hard-wired. Research has supported evolutionary explanations for sex differences (Anderson & Bushman, 2002). There also has been research attention devoted to determining the magnitude of sex differences across different ages and different types of antisocial behaviour. Research has shown that males are more likely to behave aggressively at any age. Nevertheless, regardless of sex, the effect of age is dramatic with the highest incidence of violence occurring between the ages of 15 and 35 and particularly between 15 and 24 (Anderson & Huesmann, 2003).

Baseline levels of arousal

Regarding baseline levels of arousal, individuals with lower-than-average baseline levels of arousal seem to be at greater risk of behaving aggressively. Conduct-disordered children have been found to have lower baseline heart rates and blood pressure (Rogeness, Cepeda, Macedo, Fischer, & Harris, 1990). Similarly, children and adolescents with higher levels of antisocial behaviours were found to have lower resting heart rates (Raine, 2002). The same has been found for adults, with those displaying psychopathic behaviour also showing lower arousal as measured by Electroencephalogram or EEG (Howard, 1984) or skin conductance (Hare, 1978). Such outcomes may be due to individuals with lower levels of arousal being more likely to seek situations that will provide them with more stimulation which may also lead to greater opportunities for aggression. Alternatively, they may be less likely to avoid situations in which aggression is more likely. Learning is also argued to play a role here since research has consistently demonstrated that people rather quickly habituate to repeated exposure to aggression or violence (Anderson & Huesmann, 2003).

Although some biological factors appear to be important (e.g., sex), others such as testosterone have not shown strong effects on antisocial behaviour. It is generally agreed that biological factors influence the risk of aggression but do not determine aggression itself (Raine, Brennen, Farrington, & Mednick, 1997). Rather, interaction with environmental factors provides a more accurate understanding of the development of antisocial behaviour. To reinforce the importance of such interaction, Anderson and Huesmann (2003) have highlighted that severe aggressive and violent acts rarely occur unless multiple precipitating situational and individual difference factors co-occur. A single precipitating factor will only explain a small portion of the variance in aggressive behaviour.

Reflecting this understanding, it is accepted that biological processes influence cognitive constructs such as *knowledge structures* as well as emotional functioning and responding. Similarly, there are biological effects on learning processes such as biological preparedness (Seligman, 1970), which refers to a biologically-influenced tendency to learn certain stimulus-response associations more easily than others. For example, research has shown that children are more likely to learn fear responses to certain stimuli (e.g., snakes) rather than others (power outlets) independently of the actual risk posed by the stimuli at the current time, but logically predicted by the risks these stimuli may have posed in the species' evolutionary history (Gullone, 2000). The relevance of the preparedness construct to aggression is explained by the links between aggression and frustration, as well as pain and anger, that appear to be readily formed by humans (Berkowitz, 1993). In addition to the biological preparedness to develop certain stimulus-aggression response associations, individual difference factors, including temperament and personality traits, place certain individuals at more at risk than others.

Individual difference risk factors

Temperament

The predispositional factor that perhaps most significantly predicts the risk faced by individuals growing up in a non-optimal environment is early temperament. There has been a substantial amount of longitudinal research confirming the important role played by temperament in predicting antisocial behaviour throughout development, including the preschool period (Keenan, Shaw, Delliquadri, Giovanelli, & Walsh, 1998), childhood (Raine, Reynolds, Venables, & Mednick), and the adolescence period (Caspi et al., 1995).

Two of the leading researchers in the field of temperament (Rothbart & Bates, 1998; p. 109) have described temperament as:

constitutionally based individual differences in emotional, motor, and attentional reactivity and self regulation. Temperamental characteristics are seen to demonstrate consistency across situations, as well as relative stability over time.

Thus, temperament refers to a child's behavioural style, reflecting not so much what children do as the style of their responses. For example, some children are more uninhibited or hyperactive than others. They are also more impulsive and fearless than other children. Of note, adult personality has as its earliest manifestations childhood temperament. Further, adult antisocial personality has a strong relationship to childhood antisocial behaviour (e.g., Krueger, Caspi, Moffitt, White, & Stouthamer-Loeber, 1996). Indeed, the findings regarding the relationship between temperament and antisocial behaviour are sufficiently strong that Dodge et al. (2006) have stated "the question is no longer is there a relation, but which dimensions of temperament and personality are most strongly related to antisocial behavior?" (p. 734).

Researchers have identified different temperament styles, including (i) easy temperament, (ii) difficult temperament, and (iii) slow-to-warm up. The style of most relevance to the development of antisocial behaviour and conduct problems is difficult temperament. On the basis of work conducted by Thomas and Chess (1977), it can be estimated that around 10% of infants can be classified into this category. The difficult baby is irregular in behaviour and shows discomfort when situations change. He often cries and often displays a negative mood. He is also more likely to withdraw from new experiences and to react intensely to most environmental stimulation. A number of studies have found that difficult temperament assessed early in life is predictive of the later development of antisocial behaviour (e.g., Kingston & Prior, 1995).

Other studies have been based on more discrete dimensions of temperament as compared with the classification reported by Thomas and Chess (1977). Such studies have indicated that the strongest predictors of antisocial behaviour are fearlessness, impulsivity, low effortful control, and lack of control as well as irritability, anger, or frustration. In addition to being demonstrated in short-term follow-up studies, these findings have been confirmed in longitudinal studies that have followed children over several years. One example is the study by Raine et al. (2002), which followed a sample of 1,130 children from infancy through late childhood/early adolescence. The researchers found that fearlessness and sensation-seeking at age three predicted aggression at age eleven. Henry, Caspi, Moffitt, & Silva (2001) reported that the dimension of *lack of control* measured early in life predicted violent convictions at age 18 years. These same researchers found that this same dimension significantly discriminated between life-course persistent and adolescent-limited offenders.

Personality

Research has also focused on dimensions of individual differences in adulthood that are important for better understanding antisocial behaviour (e.g., Moffitt, Krueger, Caspi, & Fagan, 2000). Lynam and Widiger (2001) examined patterns of scores on the Five-Factor Model of personality dimensions for antisocial individuals and found that they were characterized by low scores on agreeableness and conscientiousness (*see Table 1 for a description of these personality dimensions*). Others have highlighted the affective functioning of a sub-group of antisocial individuals as being characterized by a lack of empathy and guilt as well as an interpersonal style featuring the callous use of others and narcissism, characteristics that are defining features of Psychopathy (Frick, 2002). Such individuals show a particularly severe and violent pattern of behaviour. Psychopathic behaviour will be discussed in more detail in the subsequent section.

Miller and Lynam (2001) conducted a meta-analysis of 59 studies that examined the relationship between antisocial behaviour and what have been referred to as "structural models of personality". Their meta-analysis focussed on studies that involved the four most widely used structural (or trait) models of personality. Such models are proposed to identify basic traits or building blocks of individuals' characteristic ways of thinking, feeling, and behaving. The four models examined were the Five-Factor Model (McCrae and Costa, 1990); Eysenck's PEN Model (Eysenck, 1967), Tellegen's (1985) Three-Factor Model and Cloninger's Temperament and Character Model (Cloninger et al., 1993 – comprising seven traits). See Table 1 for model details.

As noted by Miller and Lynam (2001), although the four models differ with regard to the number of factors and the methods used to determine the factors, there is a great deal of conceptual similarity among them. Importantly, the models have been empirically validated and broadly represent normative personality as opposed to pathological aspects of personality. Of the 18 personality dimensions examined, Miller and Lynam found that eight factors had moderate relationships with antisocial behaviour.

The eight factors that had moderate relationships with antisocial behaviour were as follows:

- Five-Factor Model: High *Agreeableness* – Low Antisocial Behaviour
- Five-Factor Model: High *Conscientiousness* – Low Antisocial Behaviour

- Eysenck's Model: High *Psychoticism* – High Antisocial Behaviour
- Tellegen's Model: High *Negative Emotionality* – High Antisocial Behaviour
- Tellegen's Model: High *Constraint* – Low Antisocial Behaviour
- Cloninger's Model: High *Novelty Seeking* – High Antisocial Behaviour
- Cloninger's Model: High *Self-directness* – Low Antisocial Behaviour
- Cloninger's Model: High *Cooperativeness* – Low Antisocial Behaviour

Of these eight factors, those with the strongest relationships were Psychoticism, Agreeableness, and Novelty Seeking. Moreover, close examination of the eight correlated personality factors indicates that they all essentially assess Agreeableness or Conscientiousness/ Constraint. On the basis of their results, Miller and Lynam put forth the following description of the personality characteristics that typify antisocial individuals:

> Individuals who commit crimes tend to be hostile, self-centred, spiteful, jealous, and indifferent to others (i.e., low in Agreeableness). They tend to lack ambition, motivation, and perseverance, have difficulty controlling their impulses, and hold non-traditional and unconventional values and beliefs (i.e., low in Conscientiousness). (p. 780)

Importantly, as noted by Dodge et al. (2006), the relationships found between personality dimensions and antisocial behaviour cannot be argued away as simply being due to content overlap in measurement instruments, since personality inventories "contain very little reference to explicitly antisocial behavior" (p. 736). Further, the relationships have been confirmed through prospective research wherein low levels of Constraint and Agreeableness at age 18 were found to predict symptoms of Antisocial Personality Disorder at age 21 years, even after already accounting for the effect of Antisocial Personality Disorder at age 18 years (Krueger, 1999; Tackett & Krueger, 2011).

Psychopathy and Callous-Unemotional traits

One individual difference variable that is more descriptive of abnormal or disordered functioning, as compared to normative personality dimensions such as Neuroticism, is Psychopathy. Of note, Psychopathy can be conceptualized as a particular constellation of personality

Table 1 Models of personality and their dimensions.

Model	Dimension Definitions
The Five-Factor Model	
Neuroticism	Emotionally unstable, sensitive, nervous, a tendency to experience negative emotions such as anger, anxiety or depression.
Extraversion	Energetic, outgoing, a tendency to seek the company of others and to experience positive emotions.
Openness to Experience	High appreciation of art and intellectual curiosity, willingness to experience and consider new ideas, beliefs, and experiences.
Agreeableness	A tendency toward being compassionate, kind, and cooperative toward others.
Conscientiousness	Ability to show self-discipline and restraint, to control impulses, good organizational skills, tends to plan behaviour rather than to act spontaneously, follows moral code.
Eysenck's Three-Factor Model	
Psychoticism	Solitary, egocentric, cruel and inhumane, disconnected, lacking empathy, impulsive and troublesome, "lacking in feelings for his fellow-beings and for animals" (Eysenck & Eysenck, 1975; p. 1).
Extraversion	Sociable, craves excitement, takes chances, tends toward spontaneity and impulsivity, not always reliable.
Neuroticism	Emotionally unstable and maladjusted, tends to experience negative emotions including anxiety. Worries and can be moody.
Tellegen's Model	
Positive Emotionality	Well-being, sociability, achievement orientation, social closeness, tendency to experience positive emotions.
Negative Emotionality	Low stress tolerance, aggression, and alienation. Tendency to experience negative emotions.
Constraint	Harm avoidance, ability to control impulses, endorses traditional values and standards.
Cloninger's Model	
Novelty Seeking	Tendency to experience reward in response to novel stimuli.
Harm Avoidance	Avoidance of aversive stimuli. Tendency to respond intensely to such stimuli.
Reward Dependence	Tendency to respond intensely to reward signals.
Persistence	Tendency to persevere even if frustrated and fatigued.
Self-Directedness	Self-determination and willpower.
Cooperativeness	Tendency to respond or behave in an agreeable as opposed to antagonistic and hostile manner.
Self-transcendence	Experiencing of spiritual ideas and involvement in spirituality.

Source: Adapted from Miller & Lynam (2001).

traits, including low Agreeableness and low Conscientiousness. Other characteristic traits include high levels of angry hostility and impulsiveness combined with low levels of anxiety, self-consciousness, and vulnerability. Psychopaths also present with high assertiveness and excitement-seeking but low warmth (Hare, 2003).

In his article discussing the early identification of the Psychopath, Lynam (1996) noted that in any given population, the majority of offences are committed by a minority of offenders, with the most persistent 5% to 6% of offenders being responsible for 50% to 60% of known crimes. Lynam referred to such individuals as "chronic offenders". Characteristically, such individuals begin their criminal careers earlier than other offenders and continue longer. He went on to say that the most likely candidate for a criminal career is the Psychopath, which, according to Lynam is easily identified in adulthood.

Lynam (1996) described the Psychopath's characteristics according to three major domains of functioning including behavioural, interpersonal, and affective or emotional. Accordingly, within the behavioural domain, the Psychopath presents as a risk-taking sensation-seeker who is involved in a variety of criminal activities. In the interpersonal domain, the Psychopath is characterized by grandiosity, egocentrism, and manipulative, forceful, and cold-hearted actions. Within the affective domain, the Psychopath displays shallow emotions; is unable to maintain close-relationships; and lacks empathy, anxiety, and remorse. This is consistent with the work of Frick, O'Brien, Wootton, & McBurnett (1994), who highlighted that clinical reports spanning several decades document descriptions of the psychopathic personality as being characterized by egocentricity, absence of empathy and guilt, superficial charm, shallow emotions, and absence of anxiety, in addition to deviant social relationships.

Given such characteristics, it is not surprising that psychopathic offenders are not only the most persistent but also the most violent. According to Lynam (1996), "The psychopathic offender commits more crimes than the average criminal offender" and "more types of crimes, as well as more crimes of any type, relative to non-psychopathic offenders" (p. 209). The Psychopathy construct has been found to be useful in identifying a particularly severe and violent group of antisocial adults with distinct causal processes that appear to have led to their antisocial behaviour (Bair, Peschardt, Budhani, Mitchell, & Pine, 2006). Thus, research evidence suggests a unique aetiology for such offenders.

Although the characteristics of Psychopathy overlap significantly with those of Antisocial Personality Disorder, and at one time psychopathic traits and Antisocial Personality Disorder were considered to be analogous, later work (e.g., Hare, Hart, & Harpur, 1991; Harpur, Hare, & Hakstian, 1989) indicated that the antisocial behaviours associated with Antisocial Personality Disorder and the motivational and interpersonal processes associated with Psychopathy are distinct. In addition, the correlates of the two types of disorder differ. Correlates of Antisocial Personality Disorder include adverse family background factors, such as low socioeconomic status and low intelligence. In contrast, correlates of psychopathic traits include narcissism and low anxiety. Moreover, those individuals who present with both Antisocial Personality Disorder and Psychopathy features show a more severe and chronic pattern of antisocial behaviour (Frick et al., 1994).

Psychopathy in non-adult populations

More recently, research has examined the construct in non-adult populations (i.e., Frick, et al., 1994; Lynam, 1997), resulting in a burst of research attention into juvenile Psychopathy that has yielded findings highly consistent with those in adulthood (Dodge et al., 2006). The major relevant finding relating to juvenile Psychopathy is one of predictive relationships with antisocial behaviour. For example, Lynam (1997) reported that juvenile Psychopathy significantly predicted future antisocial behaviour.

It is noteworthy that Callous-Unemotional traits form a prominent part of Psychopathy conceptualizations in adults (Cleckley, 1976; Hare, 1993). Individuals characterized by Callous-Unemotional traits lack a sense of guilt and empathy and callously use others for their own gain (Frick & White, 2008). At least three dimensions have consistently emerged in the conceptualization of psychopathy in adults. They are (i) Callous-Unemotional traits, (ii) an interpersonal style characterized by arrogance as well as deceitful and manipulative behaviour, and (iii) an impulsive and irresponsible behavioural style that includes poor planning and a tendency toward boredom or need for stimulation.

Research with children has yielded three dimensions similar to those found in adult research (Frick & White, 2008). Of the three dimensions, the Callous-Unemotional traits dimension is the most characteristic of individuals who are high on other psychopathic traits when compared to other antisocial individuals who are not high on psychopathic traits. Thus, Callous-Unemotional traits appear to be a particularly useful dimension for differentiating among different sub-groups of antisocial individuals.

Of note, Callous-Unemotional traits have demonstrated stability from childhood to early adolescence and adulthood (Lynam, Caspi, Moffitt, Loeber, & Stouthamer-Loeber, 2007), and there is research showing the prediction of adult Psychopathy from adolescent Callous-Unemotional traits (Blonigen, Hicks, Kruger, Patrick & Iacono, 2006). Others have found support for the prediction of adult Psychopathy from Callous-Unemotional traits assessed as early as childhood, even after controlling for relevant childhood variables including conduct problems, antisocial behaviour, and other risk factors, such as economic disadvantage, past criminal offences, delinquent peers, and dysfunctional parenting (Burke, Loeber, & Lahey, 2007; Frick & Viding, 2009; Lynam, et al., 2007).

As noted by Frick and White (2008), research is increasingly indicating that a broad array of risk factors needs to be incorporated into theoretical models attempting to explain the development of antisocial behaviour. Another important goal is to differentiate sub-groups of youth who develop different patterns of such behaviour. Callous-Unemotional traits are of particular relevance here, since it has been demonstrated that it is those individuals identified as having child-onset antisocial behaviour who are more likely to display the Callous-Unemotional traits when compared with those who have adolescent-onset.

Research with antisocial youth has shown that Callous-Unemotional traits are predictive of a higher severity, and stability of aggressive and antisocial behaviour (Frick & Dickens, 2006). Youth who present with Callous-Unemotional traits are more likely to show both reactive and proactive aggression as well as more severe aggression. They also tend to be less responsive to cues of punishment but tend toward a reward-dominant style. This contrasts with antisocial youth without Callous-Unemotional traits, who tend to show less aggressive behaviour and whose behaviour tends to be reactive rather than proactive.

Moreover, youth high on Callous-Unemotional traits demonstrate disrupted processing of emotional cues and have lower levels of anxiety. For example, boys who scored high on a Psychopathy measure were significantly less able to correctly identify the facial expressions of sadness and fear, compared to boys scoring low on the Psychopathy measure (Stevens, Charman, & Blair, 2001). Dadds and colleagues (2006) found the deficit to be potentially explained by the failure of youth with psychopathic traits to attend to the eye region of other people's faces. This suggests that emotion-recognition problems may result from a lack of attention to aspects of the environment that are emotionally relevant (Marsee & Frick, 2010).

In their 2008 review of the social, cognitive, and emotional character-istics of antisocial youth with and without Callous-Unemotional traits, Frick and White summarized the major distinctions. These included that the conduct problems of youth without Callous-Unemotional traits are more strongly related to their dysfunctional parenting experiences. A second major finding related to the way that youth process emotional information with the presence of Callous-Unemotional traits predicting deficits in the processing of negative emotional stimuli, particularly signs of fear and distress in others. They also exhibit impairments in their emotional reactivity and moral as well as emotional socialization. They particularly seem to have difficulty responding to others' distress and consequently seem unable to inhibit the behaviours that cause the distress (Marsee & Frick, 2010). In the cognitive domain, the presence of Callous-Unemotional traits predicted a lowered sensitivity to punish-ment cues and a heightened expectation of positive outcomes in aggres-sive situations as well as a higher likelihood of verbal deficits. A fourth difference was that youth with Callous-Unemotional traits showed a higher tendency toward sensation and thrill-seeking and lowered neuroticism levels and trait anxiety.

In sum, research relating to Psychopathy and Callous-Unemotional traits in youth has revealed that an important sub-group of antisocial individuals can be designated. This sub-group differs in terms of the severity of their antisocial behaviour and its stability. The cognitive, emotional, and social processing of these youths also differs. Clearly, this population is one of particular interest when investigating indi-viduals responsible for the most severe of the antisocial and violent behaviours.

Chapter summary

This chapter reviews biological and individual difference risk factors for the development of antisocial behaviour. Research that has examined the heritability of antisocial behaviour is examined along with the way that such heritability interacts with other variables such as risky family environments. There are substantial sex differences in antisocial behav-iour that have been replicated in as many as 12 different cultures and across the childhood and early adulthood developmental periods.

Personality and temperament influences are also significant. Temperament styles characterized by impulsivity, fearlessness, low effortful control, lack of control, irritability, anger or frustration, and hyperactivity are risk factors for the development of antisocial behaviour. During the adult years, personality traits that are predictive of antisocial behaviour include psychoticism, negative emotionality, and novelty seeking. Conversely, factors protective against antisocial behaviour include agreeableness, conscientiousness, self-directedness, cooperativeness, and constraint.

In addition to particular constellations of normative personality traits, the Psychopathy variable, one that is most descriptive of abnormal or disordered functioning, is a strong marker of antisocial behaviour. There is considerable overlap of the Psychopathy construct and other personality traits, including low agreeableness, low conscientiousness, low anxiety, low self-consciousness, low empathy, and low warmth. Psychopathy is also characterized by excitement and novelty seeking as well as high assertiveness. It is noteworthy that the majority of crimes are committed by people who are high in Psychopathy. Moreover, crimes committed by psychopathic offenders are more likely to be violent crimes. There is also substantial overlap between Psychopathy and Antisocial Personality Disorder.

Of note, juvenile Psychopathy that is characterized predominantly by Callous-Unemotional traits, predicts the development of future antisocial behaviour. Research has shown that Callous-Unemotional traits are stable from childhood through to adulthood. The sub-group of a youth identified as displaying high levels of Callous-Unemotional traits differs in terms of the severity of their antisocial behaviour and in terms of their cognitive, emotional, and social processing. This group is of particular interest, given that these youth are most likely to engage in the more severe antisocial and violent behaviours.

Whilst biological and predispositional factors are of fundamental importance in understanding the developmental trajectories of antisocial individuals, the importance of environmental variables, which is the subject of the next chapter, cannot be ignored.

7
Environmental Risk Factors

Even though biological and individual difference variables such as temperament and personality play important roles in the development of aggressive and antisocial behaviours, there is evidence to suggest that environmental factors play just as important, if not a more important role. This claim is supported by the fact that recent marked increases in prevalence rates for violent crimes committed by juveniles, and wide variations across countries cannot be accounted for by enduring biological or individual difference characteristics (Dodge et al., 2006). Further reinforcing this point is the markedly higher firearm homicide rate in the U.S. compared to other industrialized countries. In the U.S., firearm homicide rates are 12 to 16 times higher than for the average of 25 other industrialized countries, a statistic that has been related to U.S. laws pertaining to gun ownership (Dodge et al., 2006).

Socioeconomic status

In addition to societal laws, other ecological factors found to be important in understanding the distribution of violent behaviours include socioeconomic status indicators, such as the proportion of families characterized by low income, unemployment, low education, number of single-parent households, and high residential mobility (Beyers, Bates, Pettit, & Dodge, 2003). However, problems have been raised with drawing firm conclusions about specific effects for these factors, since there may be relationships between family effects and neighbourhood effects. For example, Mayer and Jencks (1989) argued that families may self-select into particular neighbourhoods. Sampson, Raudenbush, & Earls (1997) coined the term "collective efficacy" to refer to societal neighbourhood factors found to be important, including supportive social networks, and the level of trust among neighbours.

Nevertheless, even when controlling for other community variables, poverty in the family continues to significantly predict higher rates of aggressive behaviour in children, adolescents, and adults (Dodge et al., 2006). Research has indicated that the mechanisms through which poverty has its impact include stressful life events and impaired social support systems (Guerra, Huesmann, Tolan, Van Acker, & Eron, 1995; McLoyd, 1990). Family poverty has also been implicated as influencing family relationship variables shown to be important in the development of antisocial behaviour, including harsh family discipline, low parental supervision, and compromised parent-child attachment (Dodge, Pettit, & Bates, 1994; Sampson & Laub, 1994). Family and parenting factors, along with situational factors, will be examined in subsequent sections of this chapter.

Provocation

According to Anderson and Bushman (2002), situational factors are important influencing factors and contribute to an understanding of processes leading to aggressive behaviour. *Provocation* has been identified as one of the strongest, if not the strongest, situational instigator of aggression in humans (Anderson & Huesmann, 2003). Examples of provocations include insults and other forms of verbal aggression, physical aggression, and blocking an individual's attempts to achieve a goal. Despite the ability to objectively identify behaviours that are likely to act as provocations, as discussed in more detail in the subsequent chapter, individuals differ to varying degrees with regard to what they perceive to be provocations. The same is true for opportunity, which is another important situational variable. Regarding opportunity, there is a tendency for people who have aggressive tendencies to not demonstrate the same self-regulatory inhibitions as do those without such tendencies.

Opportunity

Some situations provide good opportunities to aggress, whilst others restrict such opportunities. There are, however, situations wherein opportunity is not relevant, as is the case where the *removal of self-regulatory inhibitions* applies. Indeed, research has shown that normal inhibitory mechanisms that apply for most people seem to be overridden in people who have aggressive tendencies (see Keltner & Robinson, 1996; Staub 1989; 1998). As is evidenced by statistics of violence, the majority of people do not commit violent acts and would likely not do so even if they had the opportunity or even if there were little risk of being discovered or punished. Maintenance of moral standards, including self-image, self-standards, and a sense of self-worth, are all important factors in this regard. As pertinently stated by Anderson and Huesmann (2003), most people do not aggress, because they "cannot escape the consequences that they apply to themselves" (p. 309).

Aggressive cues and exposure to violence

Other important situational factors include the presence of *aggressive cues,* defined as objects that trigger memories related to aggression. For example, the presence of guns has been shown to increase aggressive responses in angry participants (Carlson, Marcus-Newhall, & Miller, 1990). Other research has shown that pictures of weapons and related words automatically prime aggressive thoughts (Anderson, Benjamin, & Bartholow, 1998). Exposure to violent television, video games, and movies has been shown to predict increases in aggressive or antisocial behaviour. It is proposed that the mechanism explaining this increase is the priming effect of exposure to the stimuli in such media. The process for this effect has been argued to be cognitive cueing (Anderson & Dill, 2000; Bushman, 1998). Cognitive cueing is likely to occur by increasing accessibility to aggressive thoughts. Also, frequent activation or cueing of concepts leads to them being chronically accessible, thereby more frequently priming antisocial behaviour (Bargh, Lombardi, & Higgins, 1988). Thus, repeated exposure to certain factors such as media violence predicts increases in antisocial behaviour (Huesmann & Miller, 1994; Patterson, DeBaryshe, & Ramsey, 1989).

A substantial amount of research has examined the effects of viewing media violence on aggressive and antisocial behaviour. In relation to this research, Anderson and Huesmann (2003) state, "regardless of how one studies the media/aggression link, the outcomes are the same – significant, substantial positive relations. This is true for longitudinal studies, cross-sectional correlational studies, field experiments, and laboratory experiments" (p. 309). In other words, research has yielded consistent findings regardless of the study method or study design. The effect sizes reported in these studies are large. When compared to effect sizes such as the effects of calcium intake on bone density or of condom use on HIV prevention, the effect sizes of viewing media violence are larger (Anderson & Huesmann, 2003). In addition to there being large effect sizes, the research outcomes are accepted as providing overwhelming evidence for a causal connection (Bushman & Anderson, 2001). Emphasizing the strength of this link, Anderson and Huesmann (2003) state, "The consistency of results regardless of design type, research group, media type, country of study, and control variables makes this one of the strongest research literatures in all of social and behavioral science" (p. 309).

Longitudinal studies provide the strongest evidence of the link. For example, research by Eron, Huesmann, Lefkowitz, and Walder (1972) and these researchers found that preferences for television violence at age eight predicted aggressiveness and antisocial behaviour ten years later, even when controlling for initial aggression levels. Later follow-ups to age 30 showed that seriousness of criminal arrests could be predicted from preferences for viewing television violence at age eight. Further, it was found that early aggression predicted criminality (Huesmann et al., 1984).

In a more recent study, Huesmann and colleagues examined 450 six- to ten-year-old children over a 15-year period. They found that childhood exposure to television violence predicted young adult aggression as determined by a combined score of self-report data, third-person report data, and crime records (i.e., proportion ever convicted of a crime; number of moving traffic violations) (Huesmann, Moise-Titus, Podolski, and Eron, 2003). Specifically, children's viewing of TV violence when aged between six and nine, their identification with aggressive same-sex TV characters, and their perceptions that TV violence is realistic, significantly predicted adult aggression. This was found for males and females. The more violence viewed, the stronger the identification, and the more the child thought that the violence reflected life "just like it is", the higher the adult aggression. This was true regardless of the child's initial aggression level. These researchers also investigated the hypothesis that "third" variables, such as social class, intellectual ability, parent aggression, or parenting differences explain the correlation between aggression and viewing TV violence through their significant relationships with each of the variables. They found little support for the "third" variable hypothesis. However, intellectual ability and parents' education did account for some of the effect in females. In their review, Anderson and Huesmann (2003) noted that the longitudinal studies in this area all include a large number of covariates, thereby refuting the argument that "third" variables explain the consistently demonstrated relationship between aggression and exposure to media violence.

On the basis of their findings, Huesmann, et al. (2003) concluded that "both males and females from all social strata and all levels of initial aggressiveness are placed at increased risk for developing aggressive and violent behaviour by adulthood when they view a high and steady diet of violent TV shows in early childhood" (p. 218). Similar conclusions have been drawn by Anderson et al. (2010) regarding the effects of exposure to violent video games. Their meta-analytic review indicated that there is strong evidence supporting causal links among increased aggressive behaviour, cognition, and affect and exposure to violent video games. The evidence also points to decreased empathy and prosocial behaviour from such exposure.

With regard to the short-term and long-term effects of media-violence exposure, Anderson and Huesmann (2003) argue that these are well understood. Exposure to media violence has a number of significant effects, including an increase in aggressive thoughts, desensitization to later violence exposure, and reductions in physiological arousal to subsequent violence exposure. Not surprisingly, viewing violent media has been shown to increase viewers' acceptance and endorsement of violent behaviour (Greeson & Williams, 1986; Hansen & Hansen, 1990). One important conclusion drawn from this research is that the effect of viewing television violence differs significantly, depending upon age of the viewer. Whilst there have been significant effects on aggression levels shown as a result of viewing violence during adolescence and early adulthood, not surprisingly, the effect has been found to be strongest for children compared to older individuals (Johnson, Cohen, Smailes, Kasen, & Brook, 2002).

Family and parenting factors

In addition to occurring at a distal cultural level and via the media, exposure to aggressive, antisocial, or violent behaviour can occur at the more proximal family level. In this section, the ways in which family variables, including witnessing of cruelty by significant others, contribute to the development of antisocial behaviour will be examined. On the whole, a number of family variables have been shown to be associated with aggression and antisocial behaviour (Rhee & Waldman, 2011).

The meta-analysis published by Repetti, Taylor, & Seeman (2002) constitutes a comprehensive examination of family-related factors that

Environmental Risk Factors 63

are important for the mental and physical health outcomes of offspring. These investigators examined the characteristics of what they referred to as "risky families" and found strong support for the important role played by the family environment in the development of antisocial behaviours. According to Repetti et al. (2002), the characteristics that appear most prominently as family "risk" factors include overt family conflict, particularly recurrent episodes of anger and aggression. Deficient nurturing or low warmth particularly characterized by cold, unsupportive or neglectful parenting was found to be another quality of a risky family. Families with such characteristics are risky because they leave children vulnerable to a range of disorders, both physical and mental.

On the basis of their analysis, Repetti and colleagues developed a model including the pathways of risk through which health in childhood is compromised and through which physical and mental health in later periods of development, including adolescence and adulthood, may be influenced. Emphasizing the importance of interactive processes between nature and nurture, they proposed a "cascade of risk" model. Certain vulnerabilities may be created by risky families, and genetically-based vulnerabilities may be exacerbated rather than attenuated, as would occur within a healthy or protective family environment. One main proposal put forth by the authors is that risky families create deficits in children's control and expression of their emotions and also in their social competence. They also argued that risky families lead to other disturbances (e.g., physiologic and neuroendocrine system regulation) and that such disturbances can have cumulative and long-term adverse effects.

Consistent with Repetti et al.'s conclusions, among the more prominent variables implicated in the development of externalizing or antisocial behaviours are parental negativity, inadequate monitoring of children's behaviour, and particular parental disciplinary practices, including power-assertive strategies and parenting that is harsh, inconsistent, or permissive (Burt, McGue, Krueger, & Iacono, 2005; Larsson, Viding, & Plomin, 2008; Loeber & Dishion, 1983; O'Connor, 2002). Indeed, a relationship between harsh and ineffective parental discipline and child aggressive behaviour problems has been reported in children as young as two to three years of age. Other important variables are insecure attachment relationships, particularly disorganized attachment, direct and indirect exposure to abuse or violence, and conflictive parent-child relationships (Lyons-Ruth, 1996; Simons, Paternite, & Shore, 2001). These variables are discussed below.

Attachment relationships

Bowlby (1969) proposed that the infant is endowed with an "attachment behavioural system" which ensures sufficient proximity to primary caregivers in order to promote survival. According to attachment theory, patterns of early interaction (the quality of care) form the foundation for individual differences in attachment, individual differences in the emerging self are based on patterns of caregiver-infant regulation, and early differences in caregiver-infant interactions have implications for patterns of adaptation in later development (Carlson & Sroufe, 1995).

Essentially, attachment theory describes a normative process in early development that is defined in terms of the regulation of behaviour and emotion. According to Bowlby, such regulation begins with the attachment that an infant forms with his or her primary caregiver. The attachment relationship represents a "special type of social relationship" (Bowlby, 1969; p. 376) and involves an affective or emotion-based bond between infant and caregiver that may be characterized in terms of regulation of infant emotion. Depending on the patterns of early care, a particular type of bond will develop between infant and caregiver.

Four attachment styles have been predominantly reported in the attachment literature. These are otherwise referred to as "internal working models" and, once developed, have been shown to be highly stable across the lifespan. The different styles include *secure attachment* characterized by positive interactions with a caregiver who is available and responsive when the infant is over aroused. Securely attached infants develop confidence in their ability to regulate emotion, including containing impulses when necessary, expressing feelings, and becoming emotionally invested in activity, as situationally appropriate. For securely attached individuals, negative emotions are not experienced as threatening but rather are seen as providing a communicative function. In contrast, infants who develop insecure attachment bonds do not fare as well.

Three insecure attachment styles have been referred to. These include (i) anxious-avoidant attachment, (ii) avoidant-resistant attachment, and (iii) disorganized attachment. Insecurely attached infants who form an *anxious avoidant* bond have been found to display overly rigid styles of emotion management. These are thought to be related to their experiences with their caregiver who has typically repeatedly ignored the infant or actively rejected the child's expressions of distress or attempts to gain reassurance. In later development, given their extremely distressing emotional experiences when faced with threat, these children tend to modify their attentional processes and avoid emotionally arousing situations. As these infants have internalized beliefs that only a restricted range of emotions is acceptable, their self-regulation of emotion is characterized by restrictions or distortions in experience and expression.

The formation of an *anxious-resistant* attachment style in infancy is associated with experiencing a caregiver as unpredictable or intermittently responsiveness to their infant's distress signals. Thus, infant communications of distress or anger do not predictably lead to emotion restabilization or to a sense of security in their caregiver's presence. These infants therefore remain vigilant and are also likely to heighten their expressions of distress in an effort to gain their caregiver's attention. Their experiences of being ineffective at regulating emotion lead to a sense of insecurity about separating from the caregiver and may support a self-regulatory style characterized by heightened arousal, exaggerated emotional expression, and a view of self as unworthy and/ or incompetent.

The fourth, most recently identified insecure attachment style, referred to as "insecure-disorganized" (Ainsworth & Eichberg, 1991; Main & Hesse, 1990), develops when the attachment figure is in some way frightened of, or frightening to, the child. Maltreated infants and infants of psychotic parents have been reported to be highly represented within this category (Main & Hesse, 1990).

A number of studies have examined the relationship between attachment style and childhood problem behaviours. As proposed by Simons et al., (2001), an insecure parent-child attachment bond might contribute to the development of hostile or antisocial behaviour through the beliefs and expectations upon which an insecure attachment relationship is based. Specifically, insecure attachment relationships are characterized by feelings of anger and insecurity, and perceptions of mistrust and chaos. Such characteristics are predictive of hostile attributions and beliefs of hostile intentions (Greenberg et al., 1993).

Investigations of the relationship between attachment security and aggressive/hostile behaviours have been classified within two categories – those conducted prior to and those conducted following the identification of the disorganized attachment style. Prior to the identification of the disorganized attachment style, a number of longitudinal studies reported relationships between insecure attachment and non-compliant or problem behaviours in infants, toddlers, and preschoolers. This was particularly true in relation to the anxious-avoidant attachment style (Lyons-Ruth, 1996). In their

1989 longitudinal study, Renken, Egeland, Marvinney, Mangelsdorf, & Sroufe investigated relationships between attachment style and aggressive behaviours with a sample of 191 children whose attachment style was assessed at age 12 months and again at 18 months. The children were followed up again when most were aged seven or eight. The researchers found that an anxious-avoidant attachment style during infancy was predictive of aggression and passive withdrawal among boys but not girls. In contrast, maternal hostility when the children were aged 3.5 predicted aggression in both girls and boys. This finding is consistent with earlier work indicating a significant relationship between parental emotional unavailability and child aggression (Egeland & Sroufe, 1981, Troy & Sroufe, 1987). Renken et al. (1989) argued that their findings provide further support for the importance of early experience for later aggressive behaviour. Of note, the finding of a relationship between avoidant attachment and aggression has predominantly been demonstrated with samples in high-risk contexts and has not consistently been found with samples in low-risk contexts (e.g., economically-advantaged families).

Studies conducted following the identification of the disorganized attachment style have indicated that a disorganized attachment style in infancy is predictive of highly aggressive behaviour later in childhood. There now exist a number of studies that have consistently reported a relationship between disorganized attachment and childhood aggression. In a study by Lyons-Ruth, Alpern, and Repacholi (1993), for example, it was found that preschoolers presenting with highly hostile behaviour were six times more likely to be classified as having a disorganized, as opposed to a secure, attachment style in infancy. For these children, the disorganized behaviour included a high degree of avoidance. In contrast, infants classified as having an avoidant, but not disorganized, attachment style were not more likely to be aggressive. The disorganized attachment style has also been shown to predict externalizing problems at age seven as reported by the children's teacher (Lyons-Ruth, Easterbrooks, & Davidson, 1997).

Thus, of the four identified attachment styles, it is the disorganized style that has been most strongly implicated in the development of hostile/aggressive behaviours. Nevertheless, some studies have shown that insecurely attached children, particularly those classified as having an anxious-avoidant attachment style, are more likely than others to interpret ambiguous situations as hostile (Dodge et al., 2006). On the whole, this research underlines the importance of the affective bond formed between infant and caregiver for the psychological development of the infant, particularly with regard to emotional and behavioural management and regulation.

Given that the disorganized attachment style develops during the infancy period or the first two years of life, and well before the development of the parent-child interaction coercive cycle (cf., Patterson, et al., 1984), it seems that attachment style itself is an important precursor to the later-appearing cycle (Lyons-Ruth, 1996).

Parenting practices

The evidence from both cross-sectional and longitudinal work that there are important differences in parenting and disciplining practices between children who are aggressive and antisocial and those who are not is substantial. In their meta-analysis of 47 studies, Rothbaum and Weisz (1994) examined the relationships of five parenting control practices with child externalizing behaviours. They found that the techniques appeared to be part of an acceptance-rejection dimension with affection and responsiveness being most descriptive at one end of the dimension, and rejection and coercion at the other end.

There is also evidence that the effects of parenting are moderated by the characteristics of the child (O'Connor, 2002). For example, Bates, Pettit, Dodge, & Ridge (1998) found a significant interaction between temperamentally-derived resistance behaviour and parenting behaviour such that a more difficult temperamental style was found to be less likely to predict problem behaviour when a firm, as opposed to less restrictive, parenting style was used. The literature on parenting has predominantly highlighted the importance of behaviours identified within Rothaum and Weisz's (1994) analysis, including warmth or sensitivity/responsiveness and conflict or hostility/rejection. Additionally, the degree of monitoring as well as the methods of monitoring have been found to be important (O'Connor, 2002).

Parental warmth

An increasing number of studies have reported parental warmth to be an important factor in understanding the development of antisocial behaviour (e.g., Booth, Rubin, & Rose-Krasnor, 1998) and the development of externalizing problems (Eisenberg et al., 2005). Indeed, Eisenberg and colleagues (2005) noted that one of the most consistent findings regarding the development of externalizing disorders, such as Conduct Disorder, is that low levels of parental warmth and support are predictive. For example, Stormshak et al. (2000) found that low levels of warmth and parental involvement predicted elevated levels of oppositional and non-compliant behaviours in children. Conversely, research has found that parental warmth plays a protective role against the development of antisocial behaviours. Of note, unfortunately, most research has been limited to examination of maternal warmth, so the findings are somewhat restricted to conclusions regarding maternal warmth. However, logically, the argument extends to paternal warmth.

Others have found that the lack of warmth manifesting through high levels of negative affect predicts the development of antisocial behaviour. For example, Caspi and colleagues (2004) found that, after controlling for genetic effects in a monozygotic (identical) twin sample, maternal negative affect (emotion) statements predicted children's antisocial behaviour. Conversely, parental expressions of positive emotions predicted low levels of problem behaviours and appeared to operate in a similar manner to parental warmth (Eisenberg et al., 2005). The mechanism underlying the association between parental emotion expression and child externalizing or problem behaviours is proposed to be emotion regulation.

Parental expression of negative affect appears to interfere with the development of the child's emotion regulation abilities, whilst positive affect expression has been found to cultivate better regulation competencies in children. As a consequence of being exposed to positive affect expression, children have been shown to be less likely to experience negative emotions, such as frustration or anger, and are also less likely to engage in aggressive or antisocial behaviours (Bariola, Hughes, & Gullone, 2011; Eisenberg et al., 2003; 2005).

Coercive and inconsistent parenting practices

Coercive parenting has been otherwise referred to as harsh, punitive, controlling, and authoritarian (Chang, Schwartz, Dodge, & McBride-Chang, 2003). Specific behaviours include yelling, name calling, frequent negative commands, physical threats and aggression, and overt expressions of anger. According to Chang et al. (2003), these behaviours can be summarized into coercive acts and negative emotional expressions. Inconsistency in parenting practices is another parenting characteristic that has been associated with higher levels of aggressive or oppositional behaviour in children (Stormshack et al., 2000). A number of studies have found this parenting dimension to be significant in distinguishing between boys who develop delinquent behaviours and those who do not (Glueck & Glueck, 1950; Parke & Duer, 1972). Indeed, an extensive body of research predominantly carried out by Patterson and colleagues has linked both coercive and inconsistent parenting practices with child aggressive and oppositional behaviours (Stormshak et al., 2000).

Patterson et al. (1989) developed a theory to explain how parenting contributes to the development of antisocial behaviour, which essentially involves a four-step learning process. The first step involves the intrusion of the parent (or family member) into the activities of their child. This may involve, for example, a mother reprimanding her child for playing too roughly with her toys. The second step might involve the child counterattacking. For example, she may complain about being reprimanded. In the third step, if the child receives reinforcement for her counterattacking behaviour through the stopping of her parent's reprimand, or if the parent does not follow through until compliance results, the child's behaviour is likely to increase as a result. In the fourth step, the parent is reinforced when the child stops the counterattack as a consequence of the parent not following through with expectation for compliance.

On the basis of their observations of parent-child interactions, Patterson and colleagues (1992) found that children respond in an aversive manner toward the parent's intrusion two to three times more frequently in distressed versus normal families. Moreover, mothers of aggressive boys more frequently reinforce their sons' aversive responses by not following through with their reprimand until compliance results. This contrasts with mothers of non-aggressive boys, who are more likely to follow through (Snyder & Patterson, 1995).

Physical punishment

Physical punishment as a disciplinary practice has also been examined in relation to the development of aggressive behaviour. It has been proposed that children follow a predictable progressive pattern toward the development of disruptive behaviours (Stormshak et al., 2000). This pattern begins with oppositional behaviours, such as non-compliance and talking back, and in some cases will progress to defiant acting out and aggressive behaviours, including hitting and physical fighting. Whilst punitive and inconsistent parenting has been proposed to predict the initiation of this pattern, physical punishment seems to be particularly characteristic at the more advanced levels.

This form of discipline was included amongst the childrearing characteristics proposed by Olweis (1995) to create bullies, along with parental attitudes of indifference toward the child, permissiveness of aggressive behaviour by the child, and power-assertive disciplinary strategies. Indeed, a notable amount of research has documented a link between physically aggressive parenting practices and elevated levels of child aggression in the home as well as at school (Stormshak et al., 2000). Research has also demonstrated a relationship between corporal punishment in childhood and antisocial behaviour, including criminal behaviour, in adulthood (Flynn, 1999b; Straus, 1991).

In her meta-analysis of 88 studies, Gershoff (2002) found that parental corporal punishment was associated with a range of child behaviours, including aggression, delinquency, and antisocial behaviour. According to Dodge and colleagues (2006), longitudinal studies have added support to the findings of Gershoff's meta-analysis by reporting a relationship between early punishment and later antisocial behaviour (e.g. Farrington & Hawkins, 1991). Nevertheless, Dodge et al. (2006) have added a qualifier to this conclusion based on the finding that the adverse effects of physical punishment appear to be attenuated when there is a warm parent-child relationship (e.g., Deater-Deckard & Dodge, 1997). Thus, it seems that parental warmth may protect the child from potential adverse outcomes brought about by physical punishment.

There is also evidence supporting the long-term adverse psychological outcomes associated with corporal punishment with research showing that such punishment is predictive of adult depression, alcohol dependence, and externalizing behaviours (Affifi, Brownridge, Cox, & Jitender, 2006). It is noteworthy, however, that adverse outcomes have most consistently been reported for more severe levels of corporal punishment as compared with occasional spanking, for example (Miller-Perrin, Perrin, & Kocur, 2009). It may be that the less severe forms of corporal or physical punishment are more likely to be combined with a more nurturing or warm parenting style, hence potentially buffering against the otherwise adverse effects.

Other studies have shown that the cultural context is of central importance in determining whether physical punishment has adverse effects on the child or not (e.g., Deater-Deckard, Dodge, Bates, & Pettit, 1996; Lansford, Deater-Deckard, Dodge, Bates, & Pettit, 2004). Essentially, such research has pointed to the conclusion that the meaning of the discipline event to the child is more important than the actual behaviour of the discipline event. If the behaviour is culturally normative, the effects of the behaviour are less likely to be developmentally compromising for the child.

Direct and indirect abuse effects

On the basis of her meta-analysis, Gershoff (2002) reported that corporal punishment is significantly associated with physical maltreatment and abuse. It is noteworthy, however, that there are differences between the two behaviours that go beyond differences of degree (Dodge et al., 2006). These include that physical abuse is often characterized by emotionally volatile and non-normative caregiver behaviours. The experience of parental abuse can be devastating for children, and there is considerable empirical evidence showing that it is one of the most important parenting variables for predicting the development of antisocial behaviour. Indeed, child abuse and child neglect are now commonly accepted to be factors that place abused or neglected children at increased risk of themselves becoming abusing or neglecting parents (Black, Heyman, & Smith Slep, 2001; Eron, 1987; Peterson, Gable, Doyle, and Ewugman, 1998; Widom, 1989). According to figures cited by Kaufman and Zigler (1987), approximately 30% of individuals who have experienced abuse (physical, sexual, or severe neglect) as children will abuse their own children, as compared to 5% in the general population. Other long-term adverse outcomes of direct and indirect abuse include being at greater risk of arrest for violent crime and of earlier and more chronic engagement in criminal behaviour.

On the basis of available evidence, it appears that physical abuse has a more consistent link with aggression than do neglect or emotional abuse (Margolin & Gordis, 2000). Increased risk of aggression and externalizing behaviour has also been reported in relation to sexual abuse. Though not as consistent as the link with physical abuse, sexual abuse has been linked with aggressive behaviour, delinquency, and other externalizing behaviours (Margolin & Gordis, 2000).

In one 20-year follow-up study, Luntz and Widom (1994) found that abused children had twice the likelihood of being diagnosed as having an Antisocial Personality Disorder compared to a matched control sample (on age, race, sex and family socioeconomic status). Other research has found that the long-term effects of abuse include school suspensions in late adolescence and physical violence (Dodge, Bates, Pettit, 1990; Lansford et al., 2002). In an attempt to control for possible child-related genetic effects, Jaffee, Caspi, Moffitt, & Taylor (2004) studied a large sample of twin pairs in Great Britain. They found that physical maltreatment is strongly causative of children's antisocial behaviour development but found no support for child genetic effects on maltreatment.

As summarized by Maughan and Cicchetti (2002), negative familial experiences, including exposure to interpersonal violence and displays of negative affect, interfere with children's developing ability to process and manage their emotions. The effects of direct maltreatment and exposure to interadult violence include deviations from normality in emotion expression, recognition, understanding, and communication. For example, Fergusson and Horwood (1998) reported strong relationships between children's observing of domestic violence and antisocial behaviour at a later time.

Research with three-month-old maltreated infants has shown that such infants display higher rates of fearfulness, anger, and sadness during interactions with their mothers when compared to non-abused children (Gaensbauer, Mrazek, & Harmon, 1981). The maltreated infants were also found to express a truncated range of emotions and to display negative emotions for higher durations when compared to their normative peers. For older maltreated children, when compared to non-maltreated peers, researchers have found higher rates of aggression (Shields & Cicchetti, 1998), withdrawn behaviour (Haskett & Kistner, 1991), and vigilance in response to aggressive stimuli (Rieder & Cicchetti, 1989).

In their study, Maughan and Cicchetti (2002) compared the socioemotional adjustment of 88 maltreated and 51 non-maltreated children who were otherwise demographically comparable. The children were aged between four and six years. Consistent with previous research, the mothers of the maltreated children in their sample reported more incidents of interadult verbal aggression and physical violence, compared to their non-maltreated sample. Also, the mothers of the maltreated children reported more child behaviour problems, compared to the non-maltreated children. Physical abuse and neglect independently predicted higher levels of socially problematic behaviours, including delinquent and withdrawn behaviour. Moreover, physical abuse was associated with child aggression when compared to non-maltreatment.

There is some evidence to indicate that the outcomes of abuse are moderated by other environmental factors, including socioeconomic status, leading to the conclusion that other types of disadvantage also play a part in the development of externalizing outcomes. For example, in their 1983 study, Wolfe and Mosk compared abused children with non-abused children from non-disadvantaged or distressed families and also with children from generally distressed families. They found that the abused children expressed more externalizing behaviours, compared to the non-abused children from non-distressed environments, but there was no difference between children who were abused and those from distressed family environments. Similar findings were reported by Toth and colleagues (1992). This is consistent with the findings of Repetti and colleagues that a common pathway to child dysfunction or pathology appears to be distress or dysfunction in the family system (cf., Repetti et al., 2002).

There is also evidence regarding the adverse outcomes associated with interadult violence exposure. Although some studies have failed to find a relationship between parental violence and child externalizing behaviours including aggression (e.g., Jouriles, Barling, & O'Leary, 1987), a significant number of studies have found such a link (see Margolin & Gordis, 2000 for review). Indeed, exposure has been associated with higher rates of both internalizing and externalizing symptomatology (Dutton, 2000; Katz &Gottman, 1993). Estimates indicate that children who witness domestic violence are at 40% to 60% greater risk of developing psychological problems, compared to children from non-violent homes (Graham-Bermann & Hughes, 1998). Studies have reported increased levels of child fear, distress, and concern in addition to anger and aggression in response to witnessed interadult anger (e.g., Cummings, 1987; Davies, Myers, Cummings, & Heindel, 1999). Some have also argued that the witnessing of violence by children places them at higher risk of developing post-traumatic stress disorder (Cunningham & Baker, 2004; Graham-Bermann & Levendosky, 1998).

A significant body of research has examined the role played by childhood exposure to violence in the development of peer relationships. Consistent with Attachment Theory proposals, it would be expected that children exposed to violence and abusive experiences are less likely to develop functional and secure peer relationships compared to children whose attachment experiences have been emotionally secure (Margolin & Gordis, 2000). Not only are abusive parents poor role models, they tend to be socially isolated. Hence, the child is less likely to be exposed to functional and healthy role models outside of the family (Davies & Cummings, 2006). Indeed, research has shown that children who are physically abused behave in a less prosocial and more aggressive manner when interacting with their peers (e.g., Alessandri, 1991; Haskett & Kistner, 1991; Kaufman & Cicchetti, 1989). It is therefore not surprising that they are likely to experience peer difficulties (Margolin & Gordis, 2000). Similar findings have been reported for children who are exposed to violence. For example, Graham-Berman and Levendosky (1998) reported that children who witnessed the abuse of their mothers (both physical and emotional) behaved in a more antisocial and aggressive manner toward their peers, compared to non-exposed children.

Peer relationships

There are complex relationships between antisocial behaviour and peer relationships. The quality of peer relationships and the ways in which individuals behave toward their peers is clearly highly related to the relationships individuals experience in their family environments. At the same time, peer relationships have a strong influence on ways of perceiving and behaving, particularly during the adolescent years. Thus, peer relationships play a significant role in influencing antisocial behaviour.

Investigations into the influence of peer factors on aggression and antisocial behaviour span from early childhood to adulthood. The importance of understanding this influence is reinforced by findings indicating that peer rejection predicts adverse outcomes, including adjustment problems and increased antisocial behaviour (Dodge et al., 2006). Of note, the context of the behaviour has been shown to be of particular importance. Studies of elementary-school classrooms have shown that the ratio of peers who are aggressive predicts children's aggressive tendencies and attitudes towards aggression (Kellam, Ling, Merisca, Brown, & Ialongo, 1998; Stormshak et al., 1999). Other research has found that the relationship between aggression and peer rejection is much stronger in girls, compared to boys, and is explained by the fact that aggression is less normative for girls. Moreover, during adolescence, aggression has been shown to increase in prevalence and consequently is considered to be more normative. Consistent with this increase, researchers have found that as children get older, they are less likely to be rejected for their aggressive behaviours (Wright, Giammarino, & Parad, 1986).

Nevertheless, research with children has consistently shown that aggressive behaviour predicts rejection by peers. A mixture of laboratory and correlational studies leads to the conclusion that early aggression contributes to rejection later in childhood (Dodge, 1983; Kupersmidt & Coie, 1990). Qualifiers that have been noted for this relationship include that not all aggressive acts are considered negative by peers. In particular, when aggression occurs in response to direct provocation, it tends to be positively, rather than negatively, evaluated. Furthermore, only about half of aggressive children are rejected by their peers. Differences between rejected and non-rejected children include that non-rejected children tend to be more socially skilled and more socially attentive, as well as less argumentative and disruptive (Bierman, Smoot, & Aumiller, 1993).

Whereas rejection is one of the most likely outcomes of aggression during the childhood years, during adolescence, adverse effects of behaving aggressively tend to relate to peer-group formation. Deviant peer groups not only bring antisocial youth together, they also contribute to the development of antisocial behaviour. In their 1992 study, Patterson, Reid, & Dishion found that youth involved with a deviant peer during sixth grade showed increases in their antisocial behaviour. Similarly, Simons, We, Conger, & Lorenz (1994) found that involvement with deviant peers predicted subsequent arrests. Others have found supporting findings (e.g., Keenan, Loeber, Zhang, & Stouthamer-Loeber, 1995). It appears that it is youth who are marginally deviant who are most susceptible to deviant peer influence rather than those who are highly deviant (Vitaro et al., 1999). Also, higher levels of deviance are predicted by holding attitudes and beliefs that committing delinquent acts is not wrong (Thornberry, Krohn, Lizotte, & Chard-Wierschem, 1993).

Research into the formation of gangs has shown consistent findings to that of peer-group formation. In fact, Spergel et al. (1989) found that youth who were members of a gang were three times more likely to engage in violent acts, compared to those not affiliated with a gang. These findings have been supported longitudinally (Thornberry et al., 1993).

The mechanisms of increased delinquency that result from deviant peer-group and gang memberships include reinforcement of delinquent behaviour. Moreover, mere membership of a deviant group includes self- and other-labelling which may yield self-fulfilling prophecies such that one identifies oneself as deviant and consequently engages in deviant behaviour. Having self-identified, and being identified by others as deviant, an individual can restore their sense of control and self-esteem through deviant behaviour (Kaplan & Liu, 1994). It is not the deviant behaviour per se that is rewarded but rather the overall set of values, attitudes, and behaviours that characterize the delinquent lifestyle (Baumeister & Boden, 1998; Dodge et al., 2006).

Chapter summary

The research examining environmental risk factors that relate to developmental pathways of antisocial behaviour is extensive. As reviewed in this chapter, micro- and macro-environmental factors not surprisingly play a strong part. Among the micro-environmental factors that have been shown to be of particular importance are situational factors, including provocation and opportunities to behave in an antisocial manner. Situational factors related to macro-environmental factors, such as neighbourhood effects, include exposure to violence and aggression as well as to antisocial individuals.

Macro-environmental factors play a significant role predominantly through their influence from micro-environmental factors. Thus, for example, families with low socioeconomic status tend to be over-represented with regard to single-parent status and high residential mobility. Parenting style, a particularly significant micro-environmental factor, is also influenced by family poverty.

The significant effects of media violence, including television violence, video-game violence, and, more recently, Internet violence, are large and rarely adequately acknowledged. Given that adult criminal behaviour can be predicted from childhood preferences for television violence, such exposure should be clearly legislated against. It is not incorrect to say that viewing media violence and playing violent media games significantly increase the risk that children's development will follow along on an aggressive, violent, or antisocial trajectory, particularly when such exposure is endorsed by significant adults in children's lives.

Much of the environmental-risk-factors literature points to the central role played by parenting and family environment experiences, as is detailed above in the *Family and Parenting Factors* section, and as highlighted clearly by Repetti and colleagues (2002) in their meta-analysis of research examining family-related factors that are important for the mental and physical health of offspring. Important findings include that deficient nurturing or low warmth and overt family conflict are characteristics of risky families or families that leave children vulnerable to a range of physical and mental health disorders. These findings incorporate a great deal of research. They are consistent with the work by Patterson and others on parental discipline, the work by Bowlby on the parent-child attachment bond, the work by Eisenberg and colleagues examining the effect of parent's emotion expression, the work reviewed by Gershoff looking at physical punishment effects, and research that has examined the more extreme behaviours of parental abuse.

Research has also reinforced the importance of exposure to interadult and interparent violence and abuse such that children who are exposed to such abuse are at 40% to 60% increased risk of developing psychological problems, compared to children without such exposure. These children are at increased risk of developing abnormal levels of fear, distress, and anxiety as well as anger and aggression. It is also noteworthy that exposure to violence plays a role in the quality of peer relations such that abused children or children exposed to abuse tend to behave in an aggressive manner and are more likely to be rejected by their peers.

During adolescence, the effects of behaving aggressively toward peers influences peer-group formation such that deviant youth are likely to become members of deviant peer groups, which in turn will contribute further to the development of antisocial behaviour. From such relationships, values, attitudes, and behaviours characteristic of a deviant lifestyle will develop, as will be discussed in the subsequent chapter.

8
Emotional and Cognitive Processes

Processes that are integral to normative and pathological development include emotion and cognition. Not surprisingly, therefore, much attention has been given to these processes in research examining the development of antisocial behaviour. This chapter will review this research beginning with emotion processes, giving particular attention to the construct of emotion regulation. This will be followed by a review of cognitive factors that directly relate to information processing. Included are knowledge structures, schemas and scripts, attributions, perceptions, and expectation biases. Research examining the ways in which these cognitive factors are related to the development of aggressive responses will also be examined. Particular attention will be given to moral disengagement, given its relevance to both information processing and antisocial behaviour. The chapter ends with a review of research that has examined the predictive value of aggressive and antisocial attitudes and beliefs and the role played by environmental factors, including parenting experiences and other micro-environment variables in the cultivation of aggressive and antisocial beliefs and attitudes.

Emotion processes

A number of emotion-related processes play a role that is specifically relevant to aggressive behaviour (Lemerise & Arsenio, 2000). Of particular relevance are the emotion-related competencies and strategies involved in regulating emotions. These processes are believed to mediate relations between environmental stimuli and antisocial behaviour.

Emotion regulation
Emotion regulation has been defined as the processes through which emotional awareness and experience are monitored, evaluated,

maintained, and modified (Thompson, 1994). According to Maughan and Cicchetti (2002), emotion regulation emerges via a developmental process resulting from both intrinsic features and extrinsic socio-emotional experiences which occur primarily within the context of parent-child interactions. It comprises a set of competencies to modulate affective states (Shields & Cicchetti, 1998). Included are competencies such as reframing upsetting events and self soothing. The ability to regulate emotion in a socially appropriate manner has important implications for interpersonal relationships and overall well-being (Gullone, Hughes, King, & Tonge, 2010; Halberstadt, Denham, & Dunsmore, 2001).

According to Eisenberg et al. (2005), parenting that is characterized by negative emotion expression and hostility is likely to cause children to become emotionally overaroused. This consequently interferes with children's emotion-regulation competency or skill development. Similarly, according to Chang et al. (2003), available evidence suggests a clear link between parenting styles and emotion regulation competencies. "The emotion dysregulation displayed by parents through harsh or punitive parenting affects the ability of their children to regulate their emotions" (p. 599). In contrast, children who experience warm and supportive parents are less likely to become overaroused and are consequently more likely to respond in an optimal manner to situational cues. In addition, positive parents are more likely to evoke positive emotions in their children, which may enhance children's abilities to regulate their own emotions and behaviour. Positive parents also model positive emotion and constructive ways of dealing with stress. As noted in the previous chapter, they are also more likely to form secure attachment bonds with their children. Such bonds are believed to foster regulated behaviour due to the greater psychological resources available to securely attached children for dealing with negative emotions. Children with secure attachments also have a better understanding of others' emotions (Laible & Thompson, 1998).

In support of the relationship between parenting behaviours and emotion regulation, at the extreme end of the harsh parenting continuum (i.e., abusive parenting), in their study comparing two groups of young children who had or had not been maltreated, Maughan and Cicchetti (2002) found that nearly 80% of the maltreated children exhibited dysregulated emotion. Relating to their results, they stated that "When feelings of emotional insecurity predominate, environmental stressors can easily overwhelm an individual's self-regulatory abilities, and dysregulation – typically of two forms: over- and underregulation – often results" (p. 1538).

A number of studies have provided evidence for direct relationships between childhood aggressive and externalizing behaviours and emotion regulation. For example, Chang et al., (2003) investigated both direct and indirect effects of harsh (coercive) parenting on child aggression in a sample of 325 children ranging between ages three and six. Their findings provide support for the direct effect of harsh parenting on children's aggressive behaviours through behavioural modelling and the indirect effect through emotion dysregulation. Eisenberg and colleagues (2005) provided additional empirical support for the mediating role of emotion regulation through their three-wave (each assessment being two years apart) longitudinal study involving 186 participants aged 11 to 16 at final assessment. The emotion regulation competency investigated by Eisenberg et al. (2005) was *effortful control*, described as involving a number of key abilities, including voluntarily focusing and shifting attention as well as the abilities to inhibit or initiate behaviour consistent with the demands of the situation. The authors concluded that "Observed parental warmth/positive expressivity in mid-elementary school predicted children's [effortful control] 2 years later, which in turn predicted low externalizing problems in adolescence" (p. 1067).

Cognitive factors and information processing

In addition to influencing emotion processes, not surprisingly, life experiences play an important role in the development of information processing factors. Life experiences also influence future behavioural tendencies (Dodge et al., 2006). A number of information processing factors are involved (Dodge, 2011). Included are processes of selective attention and encoding, attributions of intent, as well as response access and decision-making. Over time, through genetic and experiential factors, individuals develop neural pathways and associated perceptual schemas. These schemas, otherwise referred to as "knowledge structures", are stored in memory. Knowledge structures influence information processing and consequently guide our perceptions and behaviour. These structures continue to develop, in varying degrees, over time and across the lifespan (Anderson, 2002; Huesmann, 1988).

Knowledge structures

According to Anderson and Bushman (2002) "knowledge structures (a) develop out of experience; (b) influence perception at multiple levels, from basic visual patterns to complex behavioural sequences;

(c) can become automatized with use; (d) can contain (or are linked to) affective states, behavioural programs, and beliefs; and (e) are used to guide people's interpretations and behavioral responses to their social (and physical) environment" (p. 33). Three subtypes of knowledge are particularly relevant. These are (a) *perceptual schemas*, which are used to identify phenomena at varying levels of complexity, from simple everyday physical objects to complex social events and interactions; (b) *person schemas*, which include beliefs about a particular person or about a group of people; and (c) *behavioural scripts*, which contain information about the ways in which people behave in different circumstances.

Thus, knowledge structures influence perception at multiple levels and in complex ways. They influence judgements and behaviour, and they incorporate affect in a number of ways. For example, when a knowledge structure containing the emotion of anger is activated, anger will be experienced. They also contain links to knowledge about the specific emotion and so can act as a stimulus to experience specific emotions. Highlighting the broad-ranging role played by knowledge structures in everyday life, Anderson and Bushman (2002) note that "in a very real sense, personality is the sum of a person's knowledge structures" (p. 35) and knowledge structures influence the situations that an individual will seek out as well as those that they will avoid.

With increased use, and over time, knowledge structures tend to become automatic in their influence and thus increasingly function outside of conscious awareness (Schneider & Shiffrin, 1977; Todorov & Bargh, 2002). Also, over time knowledge structures become much more rigid and resistant to change. In relation to aggression-related knowledge structures, it is generally agreed that the hardening begins to take place at around ages eight or nine.

> Developing knowledge structures are like slowly hardening clay. Environmental experiences shape the clay. Changes are relatively easy to make at first, when the clay is soft, but later on changes become increasingly difficult. (Anderson, 2002; p. 70)

Schemas and scripts

In relation to sub-types of knowledge structures, Huesmann (1986) put forth what has been referred to as "script theory". Scripts are proposed to define situations and also to guide behaviour. Once scripts have been learned, they are available for retrieval at subsequent times as guides for behaviour. Scripts have been defined as "sets of particularly well-rehearsed, highly associated concepts in memory" (Anderson & Bushman, 2002; p. 31). They involve causal links, goals, and action plans. The processing of social cues is guided by scripts which are stored in memory and are the evolved representational product of experience. They influence selective attention to cues, the perception of stimuli, and the consequent decisions made on the basis of those perceptions. Script theory has proven useful for explaining the generalization of learning processes across different situations as well as the automization of perception-judgement-decision-behavioural processes (Anderson & Bushman, 2002).

Huesmann (1988) proposed that during the early developmental years, children acquire *memory scripts* which influence their perception of acceptable actions and their likely consequences. Another relevant construct is aggressive *schemas* which comprise information about elicitors and consequences of aggressive behaviour. Such schemas are argued to guide attention and interpretation of stimuli. Scripts and schemas significantly influence perception to the extent that once an individual has acquired aggressive scripts and schemas, their subsequent behaviour is likely to be primed by them (Rule & Ferguson, 1986). Indeed, research has shown that the most accessible social scripts for both aggressive children and adults are aggressive ones (Anderson & Huesmann, 2003). When compared to non-aggressive children, aggressive children are more likely to attend to aggressive social cues (Gouze, 1987). Aggressive children are also less likely to rely on external cues but more on their own stereotypes (Dodge & Tomlin, 1987), and they are more likely to describe their social relationships using such constructs (Stromquist & Strauman, 1991).

Shedding some light on the ways in which particular experiences can influence the development of particular information processing pathways, and consequently the selective attention to particular cues, Pollak and Tolley-Schell (2003) found that physically-abused children are more likely to selectively attend to angry faces and to demonstrate reduced attention to happy faces. Such children also demonstrate difficulty disengaging from angry faces. Of additional concern, it is not only children who are abused or who directly experience violence, who develop beliefs and scripts that support aggression and a tendency to behave violently, but also children who witness abuse or violence (Anderson & Huesmann, 2003).

Attributional, perception, and expectation biases

Another sub-set of knowledge structures includes cognitive processes that have been referred to as "hostile perception biases", which, according to Anderson and Bushman (2002), are examples of highly accessible hostility-related scripts. These scripts underlie the tendency by some individuals who are more aggression-prone to be more likely to perceive hostility, compared to someone who is not aggression-prone, even where there is no hostility. In other words, such individuals are biased toward perceiving hostility in situations that they find themselves in and have a related tendency to expect that others will behave aggressively, or with hostile intentions, toward them. Such biases inevitably compromise problem-solving and ultimately lead to aggressive responses (Crick & Dodge, 1994; Dodge & Tomlin, 1987). They have also been shown to be the product of well-developed schemas (Anderson & Huesmann, 2003).

Anderson and Bushman classify hostile attribution bias, hostile perception, and expectation biases (2002) as traits that predispose individuals to behave aggressively. One of the strongest findings of research related to these constructs is that individuals' perception that they are being provoked is a much more important factor in determining their response than actual provocation (Dodge, et al., 2006). If an individual perceives someone else's behaviour as having hostile intentions, such an attribution will increase the likelihood that they will react aggressively. Research has supported such a process throughout the lifespan, including from as early as kindergarten age (Shantz & Voydanoff, 1973) right through to adulthood (Baron, 1999). It has also been shown that attributions of hostile intent are promoted through socialization (Rule & Ferguson, 1986).

Attributions of hostile intent are more likely in ambiguous situations (Dodge et al., 2006). There is also research showing that aggressive children are more likely to attribute hostile intent even when intentions are benign rather than ambiguous (Dodge, Murphy, & Buchsbaum, 1984; Dodge, Pettit, McClaskey, Brown & Gottman, 1986). Of concern, longitudinal research has shown that hostile attribution biases predict increased aggressive behaviour over time.

Accessibility of aggressive responses

It is also largely agreed that the acquisition of hostile attribution biases results largely from learning processes.

> People learn specific aggressive behaviours, the likely outcome of such behaviours, and how and when to apply these behaviours. They learn hostile perception, attribution, and expectation biases, callous attitudes, and how to disengage or ignore normal empathic reactions that might serve as aggression inhibitors. (Anderson, 2002; p. 70)

Consistent with social learning theory, aggressive behaviours can be acquired through observation of models (Bandura, 1983). The observed models can be members of the child's family, a sub-culture, or even from the mass media. Aggressive models teach more than just the aggressive behavioural repertoire. Modelling provides information about general strategies and patterns of regularized sequences and contributes to the development of *scripts* (Huesmann, 1998). Huesmann (1988) proposed that aggressive scripts develop through attention to aggressive models. Thus, living in a violent family or in a violent neighbourhood, being a member of a violent peer group or gang, and watching violence on television are all candidates for teaching aggressive scripts.

Research with children has shown that attributions of hostile intent and experiencing of anger in response to particular events are more likely for those experiencing circumstances of pervasive and chronic violence, harm, and deprivation. Specific examples of such circumstances include experiences of abuse toward oneself or toward others in the family environment, including parents and siblings (Dodge et al., 2006). This relationship has been demonstrated in longitudinal research. For example, Dodge, Price, Bachorowski & Newman (1990) showed that a history of physical maltreatment in the first five years of life predicted hostile attributional bias in the elementary-school years. A history of insecure attachment and experiencing parental rejection has also been found to predict hostile attributional biases during the elementary-school years (Cassidy, Kirsh, Scolton, & Parke, 1996) as has the modelling of hostile attributional biases and attribution of hostile intent to others by mothers (MacBrayer, Milich, & Hundley, 2003).

A tendency to perceive situations and people as hostile underlies the development of aggressive schemas and scripts. That is, knowledge structures develop that support the perception of a need to be vigilant against hostility and aggression. It is likely that, for people who have this tendency toward perceptions of hostility, aggressive and hostile acts are seen to be the most effective means of obtaining desired outcomes and for solving problems (Bandura, 1977). It is therefore likely that self-efficacy beliefs support aggressive acts for such individuals.

Self-efficacy

According to Bandura's (1977) proposals about self-efficacy and outcome efficacy, individuals who believe that aggressive acts will lead to desired outcomes are more likely to engage in those behaviours. Aggressive behaviour is also more likely if supported by particular environments. Specifically, environments can increase the likelihood of aggressive behaviour if they promote the belief that such behaviour is normative or appropriate and that the behaviour will lead to the desired outcome. Aggressive behaviour is reinforced in societies that tolerate it and where reinforcements come in the form of social status or tangible rewards, or in the reduction of aversive consequences or experiences. Expressions of injury by the victim can also act as reinforcement for aggressive behaviour, given that the intention of the behaviour is to cause harm, hence such expression indicates that the desired outcome has been achieved. Importantly, direct reinforcement is not necessary for aggressive behaviour to be strengthened. As proposed by Social Learning Theory, vicarious reinforcement (that is, observing another being reinforced for their aggressive behaviour) is equally powerful a reinforcer (Bandura, 1983; Patterson, Littman, & Bricker, 1967).

In Social Learning Theory, self-reinforcement is also an important factor. It is argued that cognitive structures provide the self-derived standards against which behaviour is judged. Cognitive structures also provide the sub-functions for self-regulation, including self-monitoring, self-judging, and self-reacting. Thus, there is the selective observation of one's own behaviour. Behaviour produces self-reactions which are based upon the judgemental or self-evaluation function. Self-evaluation involves judging one's conduct against internal standards. Whereas favourable judgements result in self-rewarding reactions, unfavourable judgements give rise to negative self-reactions. According to Bandura (1978), self-generated consequences are involved in the self-regulation of aggressive behaviour. (One can theoretically extend this to antisocial behaviour more generally.) If an individual's internal standards are such that antisocial behaviour is acceptable, such behaviour will be readily engaged in and may even contribute to enhanced feelings of self-esteem. For such individuals, deterrents for antisocial behaviours come predominantly in the form of reprisal threats.

Self-deterring consequences for antisocial acts are most likely if an unambiguous connection is established between such acts and detrimental effects for the individual. However, even when such a connection is established, dissociation can be invoked between the censurable behaviour and negative self-evaluation. Bandura describes the mechanisms involved in this process of disconnection as disengagement mechanisms, as proposed within his Moral Disengagement Theory.

Moral disengagement

According to Bandura (1990; 1999), cognitive mechanisms are integral to living within one's moral standards. We learn to practice anticipatory self-censuring, and we learn to evaluate our actions and anticipate their likely consequences. Aggressive acts are predicted by processes of disengagement of the usual learned self-reactions. Certain mechanisms can explain why and when even people who otherwise have normal or even high moral standards sometimes behave in ways that could be considered reprehensible.

Bandura and colleagues (1996) specify four points in the behavioural process at which disengagement can occur. By "reconstructing the conduct, obscuring personal causal agency, misrepresenting or disregarding the injurious consequences of one's actions, and vilifying the recipients of maltreatment by blaming and devaluing them" (p. 364), self-sanctions can be disengaged. At the beginning of the behavioural process, at the point of action itself, Bandura (1978) states, "What is culpable can be made honourable through cognitive restructuring" (p. 24). Thus, one's antisocial behaviour can be self-evaluated as not reprehensible or even as righteous by contrasting it with even more reprehensible behaviour. Bandura refers to such reconstruction as "moral justification" wherein the behaviour is made socially and personally acceptable by being cast as serving some valued or moral purpose, such as supporting war for the goal of increased democracy. Examples of justification include "It is important for the well-being of our society", as could apply to soldiers fighting at war.

Cognitive restructuring can also involve the minimization of one's own moral violations by comparing them to more reprehensible acts. For example, an individual who is dishonest with his spouse may justify the action by comparing it to other men who are sexually unfaithful. This minimization mechanism has been proposed to be particularly easy for children who experience growing up "amidst truly flagrant inhumanities" (Dodge et al., 2006; p. 762) including, for example, child abuse, homicide, rape, or discrimination. Language is also a useful tool for the cognitive restructuring of behaviour. In particular, euphemistic language or language that is vague or less offensive has been empirically demonstrated to have disinhibitory power (e.g., see Bolinger, 1982). "Through sanitized and convoluted verbiage, destructive conduct is made benign and those who engage in it are relieved of a sense of personal agency" (Bandura et al., 1996; p. 365).

Another set of processes involved in moral disengagement includes a distortion of the relationship between one's actions and the effects of those actions. As in Milgram's (1974) obedience experiments, individuals can diffuse their own responsibility for the effects of their behaviour by deflecting it elsewhere. In the case of Milgram's experiments, responsibility was deflected onto the experimenter who assumed the role of the authority figure. When responsibility is displaced or diffused, people do not see themselves as accountable for their actions. With the removal of self-prohibiting reactions, uninhibited antisocial behaviour is more likely. Distorting or misrepresenting the consequences or effects of one's behaviour can also be involved in the weakening of self-deterring reactions. For example, efforts can be made to discredit evidence for the harm caused by one's behaviour. Also, the consequences of one's actions can be selectively attended to, so that the potential benefits are recalled but the harmful effects are forgotten (e.g., Brock & Buss, 1962).

A final set of mechanisms relates to the recipients of the harmful behaviour. Dehumanization is one such mechanism. This involves minimizing the moral value of the victim through redefinition so that personal moral standards no longer apply. War propaganda is considered a common vehicle for achieving this aim. Politicians throughout history have made use of such a mechanism. Such a process leads to the self-exoneration of one's inhumane acts. Yet another mechanism involved in disengagement is the attribution of blame for one's actions onto the victim. By blaming the victim, perpetrators are able to excuse their own actions. Pejorative stereotyping and indoctrination aid in the use of this mechanism (Bandura, 1978).

Research by Bandura and colleagues (1996) has shown that the use of moral disengagement mechanisms is related to people's rates of engagement in antisocial behaviours. Experimental research with children has demonstrated that when they are put into a position where they are required to evaluate behaviours and to consider the consequences for those behaviours, differences emerge between aggressive and non-aggressive children. Essentially, aggressive children are more likely to evaluate aggressive responses as more legitimate, compared to non-aggressive children (Erdley & Asher, 1998). Aggressive children have also been found to rate aggressive behaviours as less morally bad, compared to their non-aggressive peers, as more friendly and more globally acceptable (Crick & Werner, 1998; Deluty, 1983). In addition, aggressive children are more likely to expect more positive outcomes to follow from their aggressive acts, and fewer negative outcomes (Egan, Monson, & Perry, 1998; Quiggle, Panak, Garber, & Dodge, 1992; Perry, et al., 1986). In sum, during the disengagement process "the individual misrepresents, minimizes, and disregards the injurious effects of aggression while selectively focussing on the self-enhancing outcomes; diffuses responsibility for the outcomes of aggression; generates palliative comparisons for one's own aggression; places euphemistic labels on one's own aggression; and vilifies and dehumanizes the victim of aggression" (Dodge et al., 2006; p. 762).

Attitudes and beliefs

An additional set of predictive cognitive variables is *attitudes and beliefs*. Individuals who are aggression-prone demonstrate a tendency to hold positive attitudes toward aggressive or violent behaviour. These positive attitudes prepare aggression-prone individuals to behave in aggressive ways, and they also strengthen the aggressive tendency such that aggression-prone individuals are more likely to adopt aggressive problem-solving strategies.

Further highlighting the importance of family environment and parenting experiences for the development of aggression, this micro-environment has been strongly implicated in the cultivation of aggressive beliefs and attitudes. Other important environments include the neighbourhood and the child's school. Supporting research has shown that children who behave aggressively hold beliefs about social norms that influence their aggressive behaviour. Such beliefs are based upon perceptual biases influenced by their schemas and scripts and are learned through childhood experiences as well as through identification with reference groups.

Thus, children who observe significant adults in their lives behaving aggressively, not only learn those behaviours, as described by Bandura's learning theory, they also learn the belief or attitude that such ways of behaving are normal and acceptable. This has been referred to as "normative belief". Supporting this pathway of acquisition of normative beliefs, children's beliefs about aggression have been shown to be correlated with those of their parents (Huesmann, Eron, Lefkowitz, & Walder, 1984) as well as those of their peers (Huesmann & Guerra, 1997)

According to Eron (2001), there are three ways in which normative beliefs about aggression can influence an individual's perceptions. First, the stronger the normative belief in behaving aggressively, the more likely the individual (adult or child) will be to perceive hostility in others' behaviour. Second, normative beliefs about the acceptability of aggression are likely to enhance the retrieval of aggressive scripts. Third, once aggressive scripts for behaviour have been retrieved, they are more likely to be acted upon if acceptable attitudes toward aggression are held. Of note, research examining normative beliefs by Guerra et al. (1995) showed that the use of aggression is more strongly endorsed by the male culture than the female culture, that normative endorsement of aggression increases during the school years, and that normative beliefs about aggression correlate with aggressive behaviour.

Chapter summary

Emotion and cognition are two of the main domains of human functioning. Over the past two decades, there has been an increased recognition of the importance of emotion processes, particularly healthy emotion regulation, for psychological well-being. Not only is emotion regulation central to socially appropriate behaviour, it has a direct relationship with aggressive and externalizing behaviours. The emotion regulation competency of effortful control is of particular relevance.

As another of the main domains of functioning, cognition, and cognitive processes have received an enormous amount of research attention. The way we process information includes aspects of the environment that attract our attention and our perceptions of those aspects of the environment. Our perceptions, in turn, influence the ways in which we respond cognitively, emotionally, and behaviourally to that information. Constructs put forth to enhance our understanding of information processing include *perceptual schemas,* otherwise referred to as"knowledge structures", "scripts", and "attribution biases". Attitudes and beliefs are also important. For example, positive beliefs (otherwise referred to as "normative beliefs") related to aggression and violence strengthen aggressive tendencies. The same is true for *hostile perception biases.* Individuals who are more aggression-prone have a heightened tendency to perceive situations as hostile and consequently have a heightened tendency to respond aggressively. This is particularly true for individuals who have easier access to aggressive responses via aggressive scripts they may have acquired through the observation of media violence, aggressive behaviours in the home or neighbourhood, or aggressive peer-group membership.

In his Moral Disengagement Theory, Bandura proposed a number of mechanisms that enable individuals to engage in antisocial acts whilst leaving their self-image intact. These include cognitively reconstructing one's conduct, minimizing one's role as a causal agency, and blaming or devaluing the victim. Other mechanisms include diffusion of responsibility, distortion or misrepresentation of the effects of one's behaviour, and being selective with regard to the perceived consequence of one's behaviour such that the benefits are acknowledged and harmful outcomes are not.

The *normative beliefs* construct has received a fair amount of research attention, given that it is an important predictor of antisocial behaviour. The more an individual perceives an antisocial or aggressive behaviour to be normal or normative, the more likely it is that they will behave in that way. Holding such beliefs also increases access to aggressive script and increases the chances that an individual will attempt to solve perceived problems with an antisocial or aggressive response whilst managing to maintain a positive sense of self.

9
Aetiological Accounts of Animal Cruelty

As noted by Vaughn et al. (2009), research specifically examining the aetiology of animal cruelty is sparse. On the one hand, the development of animal cruelty has been researched by examining the developmental histories and experiences of children or adolescents who are cruel to animals (Currie, 2006; Duncan & Miller, 2002; Duncan, Thomas, & Miller, 2005; Thompson & Gullone, 2006). Other research has examined the childhood histories of pathological offenders to determine whether there are correlations between animal cruelty in childhood and pathological behaviour in adulthood (e.g., Felthous & Kellert, 1987; Merz-Perez et al., 2001).

Not surprisingly, regarding theories of acquisition, as with those related to aggression generally, those related to animal cruelty are, in large part, based on the conceptualization of the construct. As noted in Chapter 5, the two predominant theoretical proposals relating to animal cruelty are the Violence Graduation Hypothesis and the Deviance Generalization Hypothesis. A review of research relating to these two theoretical proposals follows. The final part of the chapter will review risk factors for the development of animal cruelty behaviours within the context of the antisocial behaviour risk factors reviewed in previous chapters.

Theoretical models of animal cruelty

The Violence Graduation Hypothesis
Several research studies published predominantly in the 1970s to 1990s investigated the proposal that animal cruelty in childhood is predictive of violence toward humans in adulthood. These studies (e.g., Felthous & Yudowitz, 1977; Kellert & Felthous, 1985; Merz-Perez, et al., 2001;

Ressler, Burgess, & Douglas, 1988) typically involved examination of the childhood histories of adult criminals and psychiatric patients. Their findings indicate support for a significant association between violence in adulthood and animal cruelty, including severe animal torture and killing in childhood and adolescence.

In much of their work, Felthous and Kellert (Felthous, 1980; Felthous & Kellert, 1986; Kellert & Felthous, 1985) compared the retrospective reports of aggressive and non-aggressive criminals with those of non-criminals. Their 1985 study involved results based on personal interviews with 152 criminals and non-criminals (i.e., 32 aggressive criminals, 18 moderately aggressive criminals, 50 non-aggressive criminals, and 52 non-criminals). They found that 25% of aggressive criminals reported five or more incidents of animal cruelty during their childhood, compared to less than 6% of moderately or non-aggressive criminals.

In 1987, these same researchers conducted a review of 15 controlled studies in which they examined "whether the scientific literature supports an association between a pattern of repeated, substantial cruelty to animals in childhood and later violence against people that is serious and recurrent" (Felthous & Kellert, 1987; p. 710). Many of the reviewed studies included animal cruelty as only one variable of interest, rather than a variable of central focus, and therefore did not investigate the relationship between animal cruelty and human violence as a primary focus. The review revealed that more studies failed to find an association compared to those that did. However, in discussing the outcomes, Felthous and Kellert identified factors that may have contributed to the overall pattern of findings. These factors included that many of the studies lacked a clear definition of animal cruelty whilst others provided no definition at all. This resulted in some studies including behaviours that may not be particularly symptomatic of abnormal aggression. In this regard, Felthous and Kellert argued that "*repeated acts*of animal cruelty are associated with personal violence which is serious and recurrent" (p. 715). They also pointed to the observation that more than half of the studies that did not find an association between childhood animal cruelty and later violent or deviant behaviour used a chart review method for collecting their data. In contrast, all of the studies that did find an association were based on direct interviews with their participants. Felthous and Kellert concluded: "The literature suggests an association between a pattern of cruelty to animals in childhood or adolescence and a pattern of dangerous and recurrent aggression against people at a later age" (p. 716).

Based on work such as that reviewed above, the Violence Graduation Hypothesis was proposed. According to this hypothesis, animal cruelty may be a form of rehearsal for human-directed violence. In developmental terms, it has been proposed that animal cruelty in childhood is a developmental incremental step toward violence directed at humans. The Humane Society of the United States (1997) coined the term the "First Strike" to refer to this association. In support of this proposal, animal welfare societies in particular have drawn upon cases of highly publicized serial killers who were abusive toward animals in their childhood.

Other research that has been cited as showing support for the Violence Graduation Hypothesis includes the work by Tingle, Barnard, Robbins, Newman, & Hutchinson (1986). This study compared the childhood and adolescent experiences of rapists and paedophiles to determine whether the two types of offenders are best grouped separately. Although not the focus of the study, the results showed that there were high frequencies of animal cruelty in both groups, with nearly half of the rapists and more than one-quarter of the paedophiles having harmed animals as children.

In their study involving 45 violent and 45 non-violent inmates in a maximum security prison, Merz-Perez, et al. (2001) found that violent inmates reported animal cruelty in their childhoods at a rate that was three times greater than that reported by the non-violent inmates. When looking at companion animals, as compared to cruelty toward other animals, the differences between the two groups were even greater, with 26% of the violent group reporting companion animal cruelty and 7% of the non-violent group.

In a review related to a more specifically-defined sample of 11 youth involved in nine incidents of multiple school shootings, Verlinden, Hersen, & Thomas (2000) found that of the 11 perpetrators involved, five (45%) had histories of alleged animal cruelty. Also in relation to a very specifically defined sample of convicted serial murderers, Wright and Hensley (2003) reported that out of 354 cases of serial murder, 75 (21%) had committed cruelty to animals during their childhood.

In an examination involving 261 inmates from medium- and high-security prisons, Tallichet and Hensley (2004) found support for the proposal that repeated acts of animal cruelty in childhood or adolescence are predictive of subsequent violent crime. In a replication of this study, Hensley, Tallichet, & Dutkiewicz (2009) examined survey data of 180 inmates from a medium and maximum security prison. As predicted, they found that recurrent acts of childhood animal cruelty were predictive of later recurrent acts of violence toward humans.

In other research, Gleyzer, Felthous, & Holzer (2002) compared 48 criminal defendants with a history of substantial animal cruelty and a matched sample of defendants without a history of animal cruelty, to investigate whether animal cruelty was associated with a diagnosis of Antisocial Personality Disorder in adulthood. They found support for the hypothesised relationship between a history of childhood cruelty to animals and a diagnosis of Antisocial Personality Disorder in adulthood. They found that a diagnosis of Antisocial Personality Disorder and also that the presence of antisocial personality traits were significantly more prevalent in the animal cruelty group.

The research cited as support for the Violence Graduation Hypothesis has been criticized as being methodologically limited (e.g., Beirne, 2004). The proposed limitations include that the research tends to be retrospective in nature and is primarily based on the self-reports provided by institutionalized individuals. Self-report has the problem of potentially being biased, and retrospective reporting has the limitation of probable recall error. Perhaps the most significant limitation is that the majority of studies have investigated the cruelty connection in highly aggressive and incarcerated criminals, thereby limiting the generalizability of the hypothesis to this group. Thus, there has been the tendency to ignore possible correlations between animal cruelty and other less severe forms of antisocial behaviour or criminal behaviour (c.f., Arluke et al., 1999). Researchers have called for longitudinal research that follows children through to adulthood in order to provide sound support for the hypothesis. It has also been argued that research is needed to rule out the possibility that the two forms of violence are not the result of other variables or a third shared factor (Flynn, 2011) such as antisocial traits.

In a recent study (Alys, Wilson, Clarke, & Toman, 2009), the problem of restricted sampling was addressed by comparing three groups of male adults. The first group comprised an incarcerated sample of male homicide sex offenders with a mean age of around 35 years; the second group comprised 20 male sex offender outpatients aged on average around 45 years; and the third group comprised 20 male students enrolled in an introductory psychology university course, with an average age of 35 years. Participants responded to questions about childhood or adolescent cruelty to animals and about other antisocial behaviour, including stealing, destruction of property, and cruelty to children. They also responded to questions about child abuse experiences and paternal alcoholism. The results revealed that all three groups significantly differed from each other. The homicide sex offenders committed significantly more animal cruelty in their younger years, compared to each of the other two groups. The differences were particularly marked between the homicide sexual offenders, nearly all of whom reported having been cruel to animals, and the university students, none of whom reported animal cruelty. There was also a significant difference between the non-homicide sex offenders and the university students, with the former group also reporting animal cruelty during their childhood and adolescent years.

Although not examining continuity of antisocial behaviour from childhood to adulthood, and therefore not providing a test of the Violence Graduation Hypothesis, the study by Flynn (1999b) does provide support for the continuity between engaging in animal cruelty during early development and attitudes indicating acceptance of interpersonal violence (normative beliefs) during later development. The study is based on the retrospective self-reports of childhood animal cruelty provided by 267 undergraduate students. Flynn examined whether there was a relationship between the students' retrospective reports of animal cruelty and their current attitudes toward various forms of family violence. As hypothesised, students who reported having been cruel to animals during childhood indicated significantly more favourable attitudes toward both corporal punishment and husbands hitting their wives than those who had not reported animal cruelty.

These last two studies cited (i.e., Alys, Wilson, Clarke, & Toman, 2009; Flynn, 1999b) add to the body of research investigating the relationships between animal cruelty during early development and related behaviours and attitudes during the adult years. They therefore go some way toward addressing the noted limitation of biased sampling. However, the methodology is still one based on self-report and retrospective reporting. Nevertheless, despite its critics, the Violence Graduation Hypothesis has continued to attract research interest (Hensely, Tallichet, & Dutkiewicz, 2009) with several more recent studies arguing support for the hypothesis (e.g., Merz-Perez et al., 2001; Merz-Perez & Heide, 2004; Tallichet & Hensley, 2004; Verlinden et al., 2000; Wright & Hensley, 2003).

Of note, as recommended by Felthous and Kellert (1987), recent work has highlighted the importance of assessing recurrent, rather than isolated, acts of childhood animal cruelty when examining the association between childhood animal cruelty and later acts of interpersonal violence. Such a position is consistent with the definition of Conduct Disorder given in the Diagnostic and Statistical Manual of Mental

Disorders. In the manual, it is stated that there must be a repetitive and persistent pattern of at least one criterion behaviour from those listed (one of which is "has been physically cruel to animals") over a period of six months. This way of defining is also in line with research findings relating to childhood-onset versus adolescent-onset of antisocial behaviour. As noted in Chapter 4, individuals in the childhood-onset group present with the more severe forms of antisocial behaviour. Indeed, consistent with claims made by the Violence Graduation Hypothesis, most violent individuals with a childhood-onset of antisocial behaviour have a developmental history characterized by an escalation in the severity of aggression (e.g., Farrington, 1991; Loeber & Hay, 1997).

Moreover, it is within the child-onset group that children with Callous-Unemotional traits are generally classified. Particularly noteworthy is the finding that Callous-Unemotional traits in childhood are predictive of later antisocial behaviour and Psychopathy (Blonigen, et al., 2006; Frick & Viding, 2009). As noted in Chapter 6, the psychopathic offender commits more crimes than the average criminal offender as well as more types of crimes. Moreover, the Psychopathy construct is useful in identifying a particularly severe and violent group of antisocial adults (Blair, Peschardt, Budhani, Mitchell, & Pine, 2006). Moreover, as previously noted, the childhood-onset group has been referred to as the "life-course persistent group".

Given that the majority of studies claiming support for the Violence Graduation Hypothesis have been based on institutionalized individuals who have committed aggressive or violent crimes, it is most likely that these individuals can best be classified as being on the life-course persistent trajectory. It is therefore reasonable to argue that the majority of Violence Graduation Hypothesis studies are reporting the same pattern of life-course persistent aggression and escalation typical of the severe end of the antisocial spectrum as has been consistently documented in the broader antisocial behaviour literature (Farrington, 1991; Loeber & Hay, 1997).

Also of note, Frick et al. (1993) found that along with fighting (first appearing at average age of 6 years), bullying (7 years), and assaulting (7.5 years), animal cruelty (6.5 years) was one of the earliest-appearing indicators of Conduct Disorder. Most importantly, Frick and colleagues found that the "Cruelty to Animals" item was one of several items that discriminated between individuals falling onto the destructive versus non-destructive end of the Conduct Disorder severity dimension. This study will be discussed in more detail in the next section. However, of relevance to the current discussion, research outside of Violence

Graduation Hypothesis supports the argument that childhood animal cruelty is one of several significant markers of the development of a more aggressive or antisocial individual.

In addition to being identified as being one of the earliest markers of the development of Conduct Disorder (Frick et al., 1993), childhood animal cruelty has been found to discriminate significantly between clinical and sub-clinical conduct problem behaviours (Gelhorn et al., 2007). Children diagnosed as having Conduct Disorder who are cruel to animals have been found to have more severe conduct problems than children diagnosed with Conduct Disorder who are not cruel to animals (Luk, Staiger, Wong, & Mathai, 1999). Consistent findings were reported in a study involving 131 children aged 6 to 13, conducted by Dadds, Whiting, & Hawes (2006). The study findings suggested that cruelty to animals may be an early manifestation of antisocial behaviour shown by a sub-group of children who develop conduct problems associated with low empathy and Callous-Unemotional traits. In other words, cruelty to animals during the childhood years may be a marker for the development of more severe conduct problems.

It therefore seems reasonable to propose that the research reporting support for the Violence Graduation Hypothesis supports the broader literature on the development of antisocial behaviour at the more extreme end of the dimension, that of early-onset individuals or of individuals who demonstrate Psychopathic and Callous-Unemotional traits. As such, it can be confidently argued that the presence of a pattern of repeated animal cruelty in young children is one behavioural marker of the developmental trajectory of life-course persistent and escalating aggression. Given that animal cruelty has been identified as one of the earliest indicators of externalizing disorders including Conduct Disorder and of development of aggression along a more severe trajectory (Frick et al., 1993; Luk et al., 1999), its early identification provides an optimal opportunity for engaging preventative strategies. As such, it is particularly important for health care professionals.

The body of research relating to antisocial behaviour has also documented a heterogeneity or breadth of antisocial acts across the spectrum of antisocial behaviour, and most particularly at the severe end (Dishion et al., 1995; Farrington, 1991; Lynam, 1996). Such research is supportive of the second hypothesis that attempts to explain animal cruelty and its aetiology – the Deviance Generalization Hypothesis. Additionally, research that has *not* targeted institutionalized or aggressive sub-types of criminal offenders, but has shown a relationship between animal cruelty and other criminal behaviours, has provided support for the Deviance Generalization Hypothesis. Of particular importance, such research has indicated that the Deviance Generalization Hypothesis applies to aggressive and antisocial behaviour across the spectrum and not just at the more severe end. Thus, the Deviance Generalization Hypothesis is consistent with current thinking that aggressive behaviours constitute a subset of behaviours classified within the Antisocial Behaviour Spectrum (Frick and Viding (2009). Research relating to the Deviance Generalization Hypothesis will be reviewed in the next section.

The Deviance Generalization Hypothesis

Although the Deviance Generalization Hypothesis is a model applied predominantly to animal cruelty, it is consistent with current conceptualizations of antisocial behaviour. As discussed in Chapter 2, current thinking relating to aggressive behaviours is that they mostly occur within the context of other antisocial behaviours such as theft, burglary, property destruction, and sexual assault and other violent crimes (Hartup. 2005). Thus, the generalization proposal put forth within the Deviance Generalization Hypothesis is consistent with current thinking that there is co-occurrence between aggressive behaviour and other forms of antisocial behaviour. Of particular note is the finding, as discussed in Chapter 2, that the greater the frequency and variety of antisocial acts, the stronger the prediction that the individual is engaged in more serious forms of antisocial behaviour, including violence (Dishion et al., 1995; Farrington, 1991). In relation to animal cruelty, this is consistent with the argument that repeated acts of animal cruelty are associated with violence that is serious and recurrent (e.g., Felthous & Kellert's, 1987). It is also consistent with findings that individuals who are cruel to animals are more likely to be engaged in a variety of other crimes than those who are not (cf. Arluke et al., 1999; Gullone & Clarke, 2008).

The literature relating to the Deviance Generalization Hypothesis that is reviewed below has been organized into a number of sections. These include work related to (i) the conceptualization of Conduct Disorder, Antisocial Personality Disorder and Psychopathy, (ii) research demonstrating support for the co-occurrence between animal cruelty and other criminal behaviours, (iii) reports of relevant work conducted by the U.S. FBI, (iv) empirical evidence supporting the co-occurrence between family violence and animal cruelty, and (v) research that has examined the links between bullying and animal cruelty as well as that examining the important role played by the witnessing of aggression.

Conduct disorder, antisocial personality disorder, and psychopathy

As previously noted, diagnostic criteria for Conduct Disorder in the *Diagnostic and Statistical Manual of Mental Disorders* (third edition) (American Psychiatric Association, 1987) and subsequent revised versions include animal cruelty as one diagnostic criterion. It is particularly noteworthy that, in their meta-analysis of child conduct problem behaviours, Frick et al. (1993) reported a median age of 6.5 years for the occurrence of the first incident of animal cruelty along with other aggressive behaviours (e.g., fighting, bullying, assaulting others), thus indicating animal cruelty to appear as one of the earliest indicators of Conduct Disorder. It is listed as such in the most recent version of the *Diagnostic and Statistical Manual of Mental Disorders* (i.e., DSM-IV-Test Revised version) (American Psychiatric Association, 2000). Further, as many as 25% of children diagnosed with Conduct Disorder display cruelty to animals. As previously noted, cruelty to animals was one of several items that discriminated between their destructive/ non-destructive dichotomy, with animal cruelty falling within the destructive category (Frick et al., 1993).

Mellor, Yeow, Hidayah, Mamat, & Hapidzal (2008) conducted an investigation with 379 Malaysian children aged 6 to 12, enabling examination of the relationship between animal cruelty and disordered behaviour in another culture. As has been found in other research (e.g., Ascione, 1993; Dadds, Whiting, & Hawes, 2006; Frick et al., 1993), the results indicated that children's animal cruelty was associated with externalizing difficulties, including Conduct Disorders and hyperactivity.

In their analysis of the National Epidemiological Survey data set including a U.S. nationally representative sample of 43,093 respondents, Gelhorn, et al. (2007) found that cruelty to animals (assessed with the item "Hurt or be cruel to an animal or pet on purpose") significantly discriminated between those with clinical and sub-clinical conduct problem behaviours. Specifically, 5.5% of males in the sub-clinical group, compared to 18% of males in the Conduct Disorder group, endorsed the item of animal cruelty. As would be expected, the comparative statistics for females were lower but equally discriminating (i.e., 2.2% versus 6.2%).

Consistent findings were reported by Luk et al. (1999), in their comparison study of 141 clinic-referred children presenting with at least one definite Conduct Disorder symptom apart from animal cruelty, and a community sample of 36 children, all aged between 5 and 12. Forty children in the clinic-referred group (out of 141–28%) compared to one child from the community sample (3%) were rated as "sometimes" or "definitely" being cruel to animals. As noted earlier, children in the animal cruelty group were found to have more severe conduct problems than the comparison group and were more likely to be male.

The older children in the animal cruelty group had a highly elevated self-perception. Luk and colleagues proposed that the elevated self-worth of the children who were cruel to animals was suggestive of the presence of Callous-Unemotional traits (c.f., Frick, O'Brien, Wootton & McBurnett, 1994) given that such traits manifest as behaviour characterized by lack of guilt and empathy and superficial charm. This finding is consistent with research by Frick and Dickens (2006). In their research with antisocial youth, they found Callous-Unemotional traits to be predictive of a higher severity, and stability of aggressive and antisocial behaviour (Frick & Dickens, 2006). In contrast, antisocial youth without Callous-Unemotional traits showed less aggressive behaviour.

Evidence that Callous-Unemotional traits and Psychopathy maybe be particularly predictive of animal cruelty behaviours is convergent with findings of research that has investigated relationships between animal cruelty and other criminal behaviours. As noted in Chapter 6, according to Lynam (1996), Psychopathy is characterized by more crimes than is true for the average criminal offender and also by more types of crimes. Such findings are reflective of the criminal behaviour profiles of people who are cruel to animals. Such a profile is also reflective of particularly severe and violent antisocial adults (Blair, Peschardt, Budhani, Mitchell, & Pine, 2006).

In a controlled study aimed at identifying risk factors for abuse and interpersonal violence among an urban population, Walton-Moss, Manganello, Frye, & Campbell (2005) compared 845 women who had experienced abuse in the past two years with a control group of non-abused women from the same metropolitan area. Risk factors for the perpetration of interpersonal violence included being a high-school drop-out, being in fair or poor mental health, having a problem with drugs or alcohol, and companion animal cruelty.

In the more recent investigation by Vaughn and colleagues (2009), the correlates of lifetime animal cruelty including Conduct Disorder and other disorders, as well as socio-demographic variables, were examined. The 2001–2002 data set comprised data from a nationally representative sample of 43,093 non-institutionalized United States residents aged 18 or older. Data were collected via interview by trained interviewers using a validated interview schedule (Grant, Harford, Dawson, & Pickering, 1995).

Among the socio-demographic variables assessed, being male predicted a higher prevalence of animal cruelty as did being younger and from a lower socio-economic background. The findings showed that the prevalence of antisocial behaviours was higher among those with a lifetime history of animal cruelty, compared to those without such a history. The most prevalent antisocial behaviours among those who were cruel to animals were crimes including robbing or mugging. Other more prevalent behaviours were setting fires on purpose, harassing and threatening someone, and forcing someone to have sex.

In additional analyses, the data revealed that animal cruelty was uniquely associated with disorders characterized by low self-control, including lifetime alcohol use, pathological gambling, Conduct Disorder, and a number of personality disorders, including Obsessive-Compulsive Personality Disorder and Histrionic Personality Disorder. Indeed, the most common psychiatric disorders among people with a history of animal cruelty were Conduct Disorder and Antisocial Personality Disorder in addition to a family history of Antisocial Personality Disorder and lifetime nicotine dependence, as well as the disorder related to lifetime alcohol use. (See Table 2 for descriptions of these disorders.) Supporting the role played by developmental family experiences, animal cruelty was also associated with a family history of antisocial behaviour.

Criminal behaviour and animal cruelty

Consistent with the research reported above that showed a correlation between criminal behaviours and animal cruelty (Vaughn et al., 2009), research by Coston and Protz (1998) demonstrated significant overlap between animal cruelty and other criminal behaviour. These researchers sought to examine the overlap by cross-referencing cases of individuals in a county in North Carolina who had been investigated for animal cruelty in 1996 with 911 calls two years earlier and one year later. They found 1,016 matches for crimes investigated two years earlier than 1996. The resulting reports were for sexual assaults (40%), mental health requests (23%), assaults (22%), animal cruelty (6%), missing person (5%), and domestic violence (4%). One-third had been arrested for criminal offences other than animal cruelty during this earlier period.

The number of matches one year later was 754, and the reports related to creating a disturbance (32%), domestic violence (31%), assault (16%), missing person (6%), man with a gun (5%), animal cruelty (4%), mental health (2%), sexual assault (2%), and drugs (1%). One-third had been arrested, and 10% had been convicted for assault, domestic violence, and drug possession.

Although the aim of their study was to examine the Violence Graduation Hypothesis, Arluke et al. (1999) found support for the Deviance Generalization Hypothesis. As with the studies described above, they investigated the relationship between criminality and animal cruelty. Their investigation included examination of other forms of antisocial behaviour, including violence. In an attempt to overcome some of the limitations of past research, they obtained their data from official records of criminality rather than through self-report from institutionalized individuals. They also included a non-criminal comparison group. Their method included identifying adults who had been prosecuted for at least one incident of animal cruelty between 1975 and 1986, and their data were extracted from the records of

Table 2 Descriptions of the disorders associated with animal cruelty

Psychological Disorder	Description
Conduct Disorder	Diagnosed in childhood, this disorder is characterized by a repetitive and persistent pattern of behaviour that disregards the basic rights of others. Major age-appropriate norms are violated. The characteristic behaviours are antisocial behaviours. This disorder is seen to be the childhood counterpart of Antisocial Personality Disorder.
Histrionic PD	A personality disorder that is characterized by excessive emotionality and attention-seeking behaviour. Need for approval is excessive as are inappropriate behaviours, including flirtatious and seductive behaviours. Behaviours can also be lively, dramatic, vivacious, and enthusiastic. These people can be self-indulgent, egocentric, and manipulative.
Obsessive-Compulsive PD	A personality disorder that is characterized by a pervasive pattern of preoccupation with orderliness, perfectionism, and mental and interpersonal control. Flexibility, openness, and efficiency are compromised.
Pathological Gambling	Classified as an impulse control disorder (i.e., an addiction) and is particularly similar to substance addictions.

the Massachusetts Society for the Prevention of Cruelty to Animals (MSPCA).

They defined and identified cruelty as cases "where an animal has been intentionally harmed physically (e.g., beaten, stabbed, shot, hanged, drowned, stoned, burned, strangled, driven over, or thrown)" (p. 966). This resulted in the identification of 153 participants of whom 146 were male. The sample had a mean age of 31 years, 58% of whom were aged younger than 21. With regard to the demographics of the abused animals, the largest proportion was dogs (69%), followed by cats (22%), and the remaining were birds, wildlife, horses, or farm animals. A control group was constituted from individuals matched to the animal cruelty group on sex, socio-economic status, age, and street of residence in the same year as the cruelty incident. The details for the control group were obtained from municipal voting lists. Following this, computerized criminal records were used to track criminal records from the state's criminal justice records system. This was done for both groups. Criminal offences were classified into five groups as (i) violent, (ii) property-related, (iii) drug-related, (iv) public disorder, and (v) other.

The study results indicated that animal abusers were significantly more likely than non-animal abusers to be involved in other forms of criminal behaviour, including violent offences. Specifically, 70% of those who were cruel to animals also committed at least one other offence, compared with 22% of the control group participants. The differences ranged from 11% for the control group and 44% for the abusive group on property-related crimes to 12% for the control group and 37% for the abusive group on public disorder-related crimes. For violent crimes, the two groups differed substantially (7% and 37% for the control and abusive groups, respectively).

Based on their findings, the authors concluded that animal cruelty appears to be one of many antisocial behaviours displayed by individuals ranging from property to personal crimes. Significant is the fact that this research study included a non-institutionalized sample of people who were cruel to animals. Thus, the finding that a single known act of animal cruelty was predictive of participation in other criminal offences is particularly compelling.

In attempt to examine whether support for Arluke et al.'s findings of deviance generalization could be replicated, I (Gullone) obtained data from the Statistical Services Division of Victoria Police for *all recorded offences* in Victoria, Australia, for the years 1994 through 2001. (See Gullone & Clarke, 2008 for more detail.) Data for the equivalent timeframe and classified into the same categories were also separately obtained *only for* alleged animal cruelty offenders.

The data for *all alleged offenders* revealed that although offences against the person constituted a relatively small proportion of the total number of crimes at an average of 7.7% of all crimes over the eight-year period, when examining the percentage *only for the alleged animal cruelty offenders*, the percentage was markedly higher at 25% (see Table 3 below). This category of offences included such crimes as homicide, rape, assault, abduction/kidnap, and harassment. Importantly, these statistics are remarkably similar to those reported by Arluke et al. (1999) as described above.

There were also differences between *all alleged offenders* and *alleged animal cruelty offenders* for the remaining three categories, but they were not as great. Of note, the category of "offences against property" is the only one that has a higher percentage for all offenders as compared to only animal cruelty offenders (see Table 3).

From these data, it appears that there is a greater likelihood that people alleged to have been cruel to animals will engage in offences against the person, including violent crimes, when compared to all alleged offenders. They are also more likely to be involved in miscellaneous offences (i.e., "other offences") and drug-related offences ("drug offences") (see Table 3 below). The last category difference is not surprising, given the reported findings of Vaughn and colleagues (2009) which showed that among the most prevalent antisocial behaviours of those who were cruel to animals were "lifetime nicotine dependence" as well as "lifetime alcohol use disorder".

Of note, when broken down by age and sex, the data across the different crime categories showed that for *all alleged offenders*, as well as for *only animal cruelty offenders*, the offenders were characteristically male. The most frequent ages for all alleged offenders during the years recorded were between 12 and 35, for both males and females,

Table 3   Comparative percentages of crimes by category for all alleged offenders and only alleged animal cruelty offenders based on Victoria Police data for the years 1994 to 2001

Category of Offence	All Offenders	Animal Cruelty Offenders
Offences against the person	7.7	25.0
Offences against property	80.7	48.4
Drug offences	3.8	6.7
Other offences	7.7	19.8

but particularly for males. For both males and females, the peak ages were between 18 and 25. When examining age and sex trends only for alleged animal cruelty offenders, the same peak in frequency between the ages of 18 and 25 was found, for both males and females. Thus, as is consistent with reported findings in the broader antisocial behaviour literature, there were markedly more males identified among all alleged offenders and also among alleged animal cruelty offenders only. Males were also overrepresented across all age categories when looking at the whole database and also at only the animal cruelty database. The particular importance of these statistics is their demonstration of demographic similarity between adults who engage in criminal behaviours of various types, particularly violent behaviours, and adults who are cruel to animals. Such data provide support not only for a link between antisocial behaviour (particularly aggression) directed at other people and that directed at animals but also for the proposal that animal cruelty can be most usefully conceptualized within a human aggression/antisocial behaviour framework.

On the basis of the above work, it can be concluded that there is substantial empirical evidence to support the proposal that animal cruelty co-occurs with other antisocial or criminal behaviours. This finding is consistent with evidence, as discussed in Chapter Two, that aggressive behaviours mostly occur within the context of other antisocial behaviours, including lying, stealing, destruction of property, burglary, sexual assault, and other violent crimes (Hartup, 2005). Reflecting support for co-occurrence at the more extreme end of the antisocial behaviour continuum is the work by the U.S. FBI. This work can also be argued to provide support for the Violence Graduation Hypothesis.

Federal Bureau of Investigation work

Special Agent Alan Brantley of the FBI's Investigative Support Unit has presented findings of the work conducted by the Investigative Support Unit at international conferences (e.g., Brantley, September, 2007). Brantley, now retired, served as a psychologist at a maximum-security prison in North Carolina before joining the FBI. Throughout the course of his work, he interviewed and profiled numerous violent criminals. In an interview conducted by a representative from the Humane Society of the United States (Lockwood & Church, 1996), Brantley stated that animal cruelty is prominent in the histories of people who are habitually violent. Their histories also reveal violence toward other children and adults as well as the destruction of property.

According to Brantley, the connection between cruelty to animals and aggression against humans was first acknowledged by the FBI in the late 1970s when 36 multiple murderers were interviewed in prison. Thirty-six per cent of them described killing and torturing animals as children, and 46% described doing so during adolescence. Brantley stated, "We believe that the real figure was much higher, but that people may not have been willing to admit it...within prisons, criminals usually don't want to talk about what they have done to animals or children for fear that other inmates may retaliate against them or that they may lose status among peers" (p. 28). Brantley (2007) provided a checklist of risk indicators for future violence. These are listed in Table 4.

According to the FBI, the definition of a mass murderer is an individual who kills at least three people on the same day. In contrast, a serial killer is someone who kills three or more victims over time in separate incidents and in a premeditated fashion (Keeney & Heide, 1995). Cited as support for the Violence Graduation Hypothesis are a number of high-profile mass and serial murder cases (Petersen & Farrington, 2007). These cases also provide support for the potentially damaging role of dysfunctional environments for the child's development. In addition, they demonstrate the generalization and heterogeneity of aggressive and antisocial behaviours.

In the United States, two well-known cases of mass murder are those committed by two high-school boys – Kip Kinkel and Luke Woodham. In May 1998, at the age of 15, Kip Kinkel opened fire on his classmates in a school shooting at Thurston High School in Springfield, Oregon. Two students were killed, and 22 others were injured. It was later discovered that he had also shot dead his parents in the family home. According to Kinkel's friends and family, he had a history of abusing and torturing animals. He had boasted about blowing up a cow and killing cats, squirrels, and chipmunks by putting lit firecrackers in their mouths (People for the Ethical Treatment of Animals; PETA, 2003).

On October 1, 1997 at Pearl High School in Pearl, Mississippi, 16-year-old Luke Woodham killed two students and injured seven others. Five months earlier, he had killed his own dog. He gave the details of the gruesome killing in his diary. These included that he and his friends had tied the dog up in a plastic bag, taken her into the woods, and listened to her howl as they beat her. They then covered the bag in lighter fluid and set it on fire with a match. His other killings involved stabbing his mother and shooting dead his ex-girlfriend and another girl (Ascione, 1999).

Table 4 Checklist of risk behaviours predictive of future violence (Brantley, September, 2007)

- Anger/low frustration tolerance
- Poor control over hostility and aggression
- Impulsivity
- History of threatening violence toward self and others
- History of violence towards self and others
- Lack of insight; unstable employment record
- Residential mobility
- Blaming of others
- Projecting of blame
- Viewing of self as a victim
- Expressions of a sense of entitlement
- Emotional lability/depression
- Quick temper; irritability; humourlessness
- Expressions of a sense of hopelessness
- Childhood abuse
- Sexual and physical abuse; maternal or paternal deprivation
- Exposure to violent role models
- Described as a loner
- Isolated/withdrawn behaviour
- Poor interpersonal relationships
- Hypersensitive to criticism
- Suspicion; fearfulness; grandiosity; injustice collecting
- Altered consciousness
- Dissociation; depersonalization
- Preoccupation with violence themes, violent movies, books, TV shows, video games and other media
- Institutionalization
- Mental health problems
- Arrest history
- Physical problems
- Head trauma; neurological problems; congenital defects
- Early adjustment problems
- Animal cruelty; school problems; fire setting; enuresis; risk taking
- Juvenile delinquency
- Odd/bizarre beliefs and behaviours
- Superstitious or magical thinking; violence fantasies; paraphilias; religious or political beliefs that justify violence, delusions
- Chemical abuse
- Alcohol abuse

On April 20, 1999, the highly publicized Columbine High School shootings were carried out by Eric Harris and Dylan Klebold. They killed 12 students and a teacher and injured more than 20 other people before they killed themselves. Both were known to brag about mutilating animals (PETA, 2003).

In April, 1996, Martin Bryant was responsible for one of Australia's most terrible mass murders. He killed a total of 35 people in a 19-hour rampage in Port Arthur, Tasmania. In fact, his crime impacted the Australian population so strongly that it resulted in the passing of a law banning gun ownership in Australia. Martin Bryant had been referred to mental health services when he was only seven years old. At age 11, he was found to have tortured and harassed animals. He was also described as tormenting his baby sister.

Animal cruelty is also highly prominent in the histories of serial killers (Lockwood & Hodge, 1998). Edmund Kemper was convicted in 1973 on eight counts of first-degree murder, having killed eight women, including his own mother. It was revealed during his trial that he had a history of animal cruelty.

Albert De Salvo, who self-confessed to be the "Boston Strangler", killed 13 women between 1962 and 1963. He was also charged with armed robbery, assault, and sex offences. In his youth, he had trapped dogs and cats in orange crates and shot arrows through the boxes.

Born in 1938, Carroll Edward Cole was an American serial killer who was executed in 1985. Also known as Son of Sam, David Berkowitz pleaded guilty to 13 murder and attempted murder charges. He terrorized New York City from July 1976 until his arrest in August 1977. He was also known to have shot a neighbour's Labrador retriever. According to Lockwood and Hodge's (1998) article, Carroll Edward Cole, also regarded as "one of the most prolific killers in modern history" (p. 78) was convicted for 5 of the 35 murders for which he was accused. Cole was forced by his mother to watch as she had sexual encounters with men. He was often whipped and beaten by his mother. His first act of violence was the strangling of a puppy when he was a child.

The Moors murders were carried out by Ian Brady and Myra Hindley between July 1963 and October 1965, in the United Kingdom. The victims were five children aged between 10 and 17. Ian Brady frequently threw alley cats out of apartment windows to watch them splatter on the pavement (Petersen & Farrington, 2007). Another serial killer who is regarded as one of the most depraved in history, is Henry Lee Lucas. Lucas who was born in the United States in 1936 and died in 2001. He grew up being forced by his mother to watch her having sex with strangers. His mother also beat him and took pleasure in killing his pets. He later was introduced to bestiality and animal torture by his half brother. Together they enjoyed slitting the throats of small animals and then sexually violating their corpses (Petersen & Farrington, 2007).

There are many more examples provided by others. Despite the methodological limitations of these many accounts, including that some of the descriptions are partly derived from second hand accounts, and retrospective reports, they have substantial commonalities, including a strong suggestion of the presentation of Callous-Unemotional traits and pathological behaviour. On the basis of accounts such as those highlighted above, Lockwood and Hodge (1998) conclude that "serial killers almost invariably have histories of animal abuse earlier in their lives" (p. 80).

Another area of support for the Deviance Generalization Hypothesis within the animal cruelty literature is the research reporting the comorbidity between animal cruelty and domestic or family violence.

Family violence and animal cruelty

One of the most consistently replicated findings in the animal cruelty literature is a significant co-occurrence between family or domestic violence and animal cruelty. This research has found that more than 50% of all abused women have companion animals, and in as many as 50% of cases, the animals are abused by the perpetrators of the domestic violence. Motivations for the abuse include hurting and/or controlling the women or their children. The research has also consistently found that concern for the safety of their companion animals keeps many women (and their children) from leaving or staying separated from their abusers. It can be argued that animal cruelty, when it occurs within the family home, is a symptom of a deeply dysfunctional family (Lockwood & Hodge (1986).

One of the earliest studies to investigate the relationship between family environment and animal cruelty was the U.K. study by Hutton (1983), who reported RSPCA cruelty data for a community in England. The data showed that out of 23 families with a history of animal cruelty, 82% had also been identified by human social services as having children who were at risk of abuse or neglect.

In more recent years, several studies have investigated the relationship between family violence and animal cruelty (e.g., Ascione, 1998; Ascione, et al., 2007; Daniell, 2001; Faver & Cavazos, 2007; Flynn, 2000; Quinlisk, 1999; Volant, et al., 2008). These studies have been conducted across several countries, including the United States, Canada, and Australia. Of note, the findings are remarkably consistent across the studies despite their differences in parameters such as country where the study was conducted, sample size, and methodology used. Findings include that between 11.8% and 39.4% of women have reported that the perpetrator *threatened* to hurt or kill their companion animal(s). Between 25.6 per cent (Flynn, 2000) and 79.3% (Quinklish, 1999) of women reported that the perpetrator had *actually* hurt or killed their companion animal(s). Many of the studies examining animal cruelty within abusive families have also reported that between 18% (Ascione, 1998) and 48% (Carlisle-Frank et al., 2004) of women have delayed leaving their violent situation out of fear that their companion animal(s) would be harmed or killed if they were to leave.

A limitation of these studies, with few exceptions (i.e., Ascione et al., 2007; Volant et al., 2008), is that they did not include a comparison group of women who were not in a violent family situation. In Ascione and colleague's study, 5% of non-abused women reported companion animal cruelty, and in Volant's study led by Gullone, 0% reported companion animal cruelty. This study involved a group of 102 women recruited through 24 domestic violence services in the state of Victoria and a non-domestic violence comparison group (102 women) recruited from the community. The findings included that 46% of women in the domestic violence sample reported that their partner had *threatened to* hurt or kill their companion animal(s), compared with 6% of women in the community sample.

The focus of studies examining the relationship between family violence and companion animal cruelty has predominantly been on (i) determining the prevalence of companion animal cruelty within physically violent relationships, and (ii) the prevalence of women who delay leaving their violent relationship for fear of harm befalling their pet(s) in their absence, as well as the length of the delay. A smaller number of studies have investigated motivations underlying the companion animal cruelty in the context of family violence. On the basis of these studies, it appears that the predominant motivation is one of control. For example, in his qualitative study involving 10 women seeking refuge from domestic violence, Flynn (2000) found that batterers use animal cruelty to intimidate, to hurt, or to control their partners.

However, not all batterers are cruel to animals. To determine whether batterers who are cruel to their companion animals differ from those who are not, Simmons and Lehmann (2007) investigated the reports of 1,283 female companion animal owners who were seeking refuge from partner abuse. They found that batterers who were cruel to animals (not all battered animals were companion animals) used more forms of violence than those who were not. Specifically, batterers who were cruel to companion animals had higher rates of sexual violence, marital rape, emotional violence, and stalking. They also used more control-ling behaviours, including isolation, male privilege, blaming, intimida-tion, threats, and economic abuse. The differences were even greater for those who killed a companion animal, compared to those who did not abuse animals.

A study by Loring and Bolden-Hines (2004) involved 107 women who had been emotionally and physically abused and who were referred to a family violence centre. Each of the women had committed at least one illegal act, and 72 (62%) of the women had owned companion animals in the previous year or during the year in which the study was conducted. As many as 54 (75%) of the 72 women reported actual or threatened companion animal cruelty, and of these 54 women, 24 reported that they had been coerced to commit an illegal act through threats or actual harm to their companion animal(s).

A more recent study by DeGue and DeLillo (2009), which involved 860 university students from three U.S., universities showed that around about 60% of participants who had witnessed or perpetrated animal cruelty as a child also retrospectively reported experiences of child maltreatment or domestic violence. The study results also showed that those who had been sexually or physically abused or neglected as children were those most likely to report that they had been cruel to animals as children.

Bullying and animal cruelty in youth

In addition to being linked with abusive childhood experiences, animal cruelty has been shown to co-occur with bullying behaviours. Reinforcing their link, both animal cruelty and bullying have been related to later antisocial behaviours and Antisocial Personality Disorder (Gelhorn, et al., 2007). Not surprisingly, there are also conceptual simi-larities between animal cruelty and bullying behaviours. These include overlapping definitional criteria. For example, bullying has been defined as behaviour that is intended to hurt the victim and that is character-ized by a power imbalance, an unjust use of power, enjoyment by the aggressor, and a general sense of being oppressed on the part of the victim (Rigby, 2002). It is generally agreed that a definition of bullying needs to include an intention to inflict either verbal, physical, or psycholog-ical harm, a victim who does not provoke the bullying behaviours, and occurrences in familiar social groups (Baldry, 1998; Baldry & Farrington, 2000; Griffin & Gross, 2004; Gumpel & Meadan, 2000).

Whilst explicit in definitions of bullying but not in definitions of animal cruelty, there is a clear power imbalance where the perpetrator is more powerful than the victim and uses this power to inflict physical, emotional, or psychological harm on the victim. Both animal cruelty and bullying behaviours are predominantly observed in male populations. Males have rates of animal cruelty that are four times higher than those of females (Flynn, 1999b). They are also more likely than females to engage in bullying behaviours (Baldry, 1998; Bosworth et al., 1999; Smith & Myron-Wilson, 1998; Veenstra et al., 2005). Further suggestive of overlapping processes between animal cruelty and bullying is their appearance within a close developmental timeframe (Frick et al., 1993). Despite this strong conceptual overlap, animal cruelty and bullying behaviours have been researched separately for the most part.

The exceptions include a study by Baldry (2005), who examined the prevalence of animal cruelty, bullying behaviours, and being a victim of bullying, in an Italian sample of children and adolescents aged 9 to 12. Her results showed that girls and boys who had engaged in direct bullying behaviours were twice as likely to have been cruel to animals, compared with their non-bullying peers. Engagement in animal cruelty by boys was predicted by their direct victimization at school and indirect bullying, while engagement in animal cruelty by girls was predicted by their exposure to animal cruelty and by their experience of verbal abuse by their fathers.

Involving a school-based sample of 249 adolescents (105 males, 144 females) ranging in age from 12 to 16 years, Gullone and Robertson (2008) investigated relationships between self-reported animal cruelty and bullying. Significant positive relationships were found between bullying and animal cruelty. Both behaviours were also found to correlate significantly with bullying victimization, witnessing of animal cruelty, and family conflict.

A 2007 study by Henry and Sanders involved 185 psychology undergraduate university males. The researchers justified their decision to include only males in their study on the basis that rates of animal cruelty are substantially lower among females. Applying a retrospective reporting methodology, the study aimed to investigate the relationships between self-reports of animal cruelty and bullying as well as being a victim of bullying. They also investigated whether the relationship varied depending upon the frequency of animal cruelty or the individual's classification of bully, victim, or bully/victim. They hypothesized that the relationship between bullying and animal cruelty would be strongest for those who had been involved in multiple acts of animal cruelty as compared with isolated acts. They also hypothesized that the relationship would be strongest for those in the bully/victim group, given research suggesting that this group has the highest level of maladjustment.

The findings indicated a marked distinction between those who had been involved in a single versus multiple acts of animal cruelty. Those who reported multiple acts of animal cruelty were more likely to be classified into the bully/victim group, compared to those involved in a single act of animal cruelty. The authors concluded that their findings support the proposal that animal cruelty may sometimes constitute displaced aggression. They also concluded that high rates of bullying and of victimization are predictive of multiple acts of animal cruelty and vice versa.

In summary, research has investigated the co-occurrence of a number of aggressive or violent behaviours. Such research has provided support for the Deviance Generalization Hypothesis. Disordered functioning characteristic of Conduct Disorder, Antisocial Personality Disorder, and Psychopathy has been found to include animal cruelty amongst other aggressive behaviours. Behaviours that are characteristic of disordered functioning have also been found to co-occur with animal cruelty. For example, children who bully are also more likely to be cruel to animals. People who commit crimes, particularly violent crimes including partner or child abuse, are more likely to be cruel to animals, compared to people who have not committed these other crimes. At the more extreme end of the antisocial behaviour continuum, FBI work has shown that animal cruelty is a prominent behaviour in the profiles of violent criminals.

In addition to the research showing the co-occurrence of deviant behaviours with animal cruelty and therefore providing support for the Deviance Generalization Hypothesis, aetiological investigations provide support for the argument that animal cruelty is an aggressive and violent behaviour that cannot logically be separated from other aggressive and violent behaviours. Research shows that animal cruelty shares many of the etiological pathways and risk factors that have been shown for other deviant and antisocial behaviours. The next section will review these risk factors and pathways.

Risk factors for the development of animal cruelty

As with the broader literature on antisocial behaviour, empirical studies examining factors that are predictive of animal cruelty include a number of constitutional or biological risk factors and individual difference risk factors for the development of abusive behaviours. Being male is a consistently demonstrated risk factor across the developmental spectrum (Arluke & Luke, 1997; Coston & Protz, 1998). Age is another important constitutional variable (Arluke & Luke, 1997; Gullone & Clarke, 2008). Environmental factors are also important. These include micro-environments that can also be referred to as "proximal environments", such as the child's family and parenting experiences (e.g., Kellert & Felthous, 1985; Rigdon & Tapia, 1977; Tapia, 1971). They also include macro-environments which are considered to be more distal environments, such as cultural attitudes and norms (Flynn, 1999a).

In his recent review, Flynn (2011) listed what he considers to be the leading antecedents of children's animal cruelty. These include "a) being a victim of physical or sexual abuse, b) witnessing violence between one's parents, and c) witnessing parents or peers harm animals" (p. 455). Other antecedents or predictors of animal cruelty that Flynn included were the experiences of being bullied or the behaviour of bullying. The research examining the proposed risk factors for the development of animal cruelty behaviours will be reviewed below, beginning with biological and maturational variables.

The recent study conducted by Vaughn and colleagues (2009) is one of the largest and most comprehensive studies to date that has investigated risk factors, which they referred to as "adversities" for the development of animal cruelty behaviours. Given the demonstrated relationship between bullying and animal cruelty, the study also included bullying as a variable of interest. The study was conducted in the United States and, as described earlier, was based on data derived from the first two waves of a national epidemiologic survey regarding alcohol and related disorders. The results showed that a number of risk factors that were included in the analysis were significant.

For bullying, the risk factors included:

• Being made to do chores that were too difficult or dangerous,
• Threatening to hit or throw something
• Pushing, shoving, slapping, or hitting
• Hitting that left bruises, marks, or injuries

For animal cruelty, the risk factors included:

• Swearing and saying hurtful things
• Having a parent or other adult living within the home who went to jail or prison
• An adult/other person fondling/touching in a sexual way

Supportive of the main tenet of this book, it is important to note that cruelty to animals was significantly associated with *all assessed* antisocial behaviours. As detailed earlier in the *Conduct Disorder, Antisocial Personality Disorder, and Psychopathy* section above, strong associations were found between animal cruelty and Lifetime Alcohol Use Disorders, Conduct Disorder, Antisocial, Obsessive–Compulsive, and Histrionic Personality Disorders, pathological gambling, and a family history of antisocial behaviour.

The researchers concluded that:

> [C]ruelty to animals is associated with elevated rates observed in young, poor, men with family histories of antisocial behavior and personal histories of conduct disorder in childhood, and antisocial, obsessive–compulsive and histrionic personality disorders, and pathological gambling in adulthood. Given these associations, and the widespread ownership of pets and animals, effective screening of children, adolescents and adults for animal cruelty and appropriate mental health interventions should be deployed.
>
> (Vaughn et al., 2009, abstract)

In addition to identifying important risk factors for animal cruelty and bullying behaviours, the findings of this study provide additional support for the Deviance Generalization Hypothesis.

Sex differences

Consistent with the broader antisocial behaviour literature showing that there are marked sex differences with the males outnumbering females on aggressive tendencies by a ratio of about 10 to 1 (Loeber & Hay, 1997), research has shown that males are more likely to be cruel to animals. This is true for childhood (e.g., Baldry, 2005), adolescence (Thompson & Gullone, 2006), and adulthood (Gullone & Clarke, 2008). Interestingly, Flynn (1999a; 1999b) found that not only were males more likely to commit animal cruelty, they were also more likely to witness it.

Investigating a childhood community sample involving 268 girls and 264 boys (aged 9 to12), Baldry (2005) found that 35.9% of girls reported abusing animals, compared with 45.7% of boys. The investigation by Thompson and Gullone (2006) involved 281 adolescents aged between 12 and 18. It was found that males scored significantly higher than females on two different self-report animal cruelty questionnaires. In the study cited above by Gullone and Robertson (2008), boys were also found to score significantly higher on measures of animal cruelty, compared to girls.

Studies examining animal cruelty in adults have also found a higher prevalence amongst men compared to women. For example, in an investigation of all animal cruelty cases prosecuted in Massachusetts between 1975 and 1996, Arluke and Luke (1997) found that approximately 97% of the perpetrators were male. Similarly, in Gullone and Clarke's (2008) report of Australian data for all recorded offences in Victoria for the years spanning 1994 to 2001 (as detailed in an earlier section of this chapter), when broken down by age and sex, the data showed that across crime categories including animal cruelty, offenders were characteristically male. They also found that males were overrepresented across all age categories but most particularly between the ages of 18 and 35.

Age differences

As has been found for other forms of violence, late adolescence and early adulthood are the ages that are most typical for perpetrating animal cruelty. For example, Arluke and Luke (1997) reported that the average age for committing animal cruelty was 30. They also found that just over one-quarter of the offenders were adolescents and more than half (56%) were under 30 years. Consistent findings were reported by Gullone and Clarke (2008) in their examination of all recorded offences in Victoria, Australia, between 1994 and 2001. As reported in a previous section, in addition to being male, most offenders for all offences were aged between 18 and 35. When looking only at animal cruelty offences, there was a peak between ages 18 and 25.

In a study of 28 convicted and incarcerated male sexual homicide perpetrators, Ressler, Burgess, & Douglas (1988) found that prevalence of cruelty to animals was 36% in childhood and 46% in adolescence. Of note, in their study, Arluke and Luke (1997) also found differences depending on age, in the type of animal abused. Adults were more likely to be cruel to dogs, whilst adolescents were more likely to kill cats. The type of cruelty also differed, with the shooting of animals being more characteristic of adult animal cruelty, and beating being more characteristic of adolescent cruelty.

The finding that there are age differences in the propensity to be cruel to animals is not surprising, given the profound differences that are associated with different developmental milestones. Not only does physical strength increase as children mature, cognitive functioning and emotion regulation also develop. Moreover, environmental experiences will vary in their intensity of impact, depending on developmental stage, as has been shown for the witnessing of cruelty, as will be discussed in the next section.

Witnessing of violence, and animal cruelty

Research has demonstrated the importance of witnessing aggression for the development of aggressive behaviour (e.g., Cummings, 1987; Davies et al., 1999; Margolin & Gordis, 2000; Maughan & Cicchetti, 2002). A number of studies investigating the relationship between animal cruelty and family violence have also obtained data relating to the children's witnessing of the animal cruelty and the children's animal cruelty. The findings have indicated that between 29% and 75% of children in violent families have witnessed the animal cruelty, and between 10% and 57% have engaged in animal cruelty. As noted by Ascione et al. (2007), parental reports of animal cruelty in normative samples of children are typically around 10% or lower. Not surprisingly, these results are consistent with other studies reporting that children exposed to domestic violence are more likely to engage in acts of animal cruelty than children who have not been exposed to domestic violence (Baldry, 2005; Becker, Stuewig, Herrera, & McCloskey, 2004; Flynn, 2000; Hensley & Tallichet, 2005).

In her 2005 study, Baldry found that youth who witnessed violence between family members, or who witnessed harm to animals, were three times more likely to be cruel to animals, compared to peers without such experiences.

Currie (2006) also reported a significant relationship between the witnessing of aggressive behaviour (domestic violence) and animal cruelty via parent-report. Mother reports regarding their children's animal cruelty were compared for a group of 94 children (47 mothers) with a history of domestic violence and 90 children (45 mothers) without a history of domestic violence. According to the mother reports, exposed children were more likely to be cruel to animals, compared to children who had not been exposed to violence. Further support for this relationship was found in the more recently published study by DeGue and DiLillo (2009). These researchers found that those participants who had witnessed animal cruelty were eight times more likely than those who had not, to perpetrate animal cruelty.

In research specifically examining the relationship between children's aggressive behaviours and their witnessing of domestic violence, Baldry (2003) found that children who engaged in bullying behaviours were 1.8 times more likely to have been exposed to domestic violence than those who were not. Similarly, in their study of 281 (113 males; 168 females) school-based adolescents ranging in age between 12 and 18, Thompson and Gullone (2006) found that those who reported the witnessing of animal cruelty on at least one occasion also reported significantly higher levels of animal cruelty, compared to youth who did not witness animal cruelty. Of particular note is Thompson and Gullone's finding that witnessing a stranger abusing an animal, predicted *lower* levels of animal cruelty. This contrasted with the finding that witnessing animal cruelty by a friend, relative, parent, or sibling predicted higher levels of cruelty. These findings support Albert Bandura's vicarious learning theory, which proposes that observation of behaviour is more likely to lead to performance of the observed behaviour if the model has a meaningful relationship with the observer, or, in other words, if the model is a significant other.

Further indicating the important aetiological role of witnessing cruelty is the study by Gullone and Robertson (2008). In this study, possible pathways of acquisition for bullying and for animal cruelty behaviours were investigated. It was found that each type of behaviour was significantly predicted by the witnessing of animal cruelty. Thus, this study not only supports the co-existence of animal-directed aggression and human-directed aggression in youth, as with Baldry's (2005) results, it also demonstrates the importance of observational learning, in this case the observation of animal cruelty, as a pathway for the development of different aggressive behaviours.

Others (e.g., Flynn, 1999b; 2000; Henry, 2004b; Hensley & Tallichet, 2005) have demonstrated this relationship by asking undergraduate students or imprisoned males about their childhood experiences and behaviours. A study by Henry (2004a) involved 169 university students who were asked about exposure to, and perpetration of, animal cruelty. Results indicated that animal cruelty had been witnessed on at least one occasion by 50.9% of participants. Furthermore, the witnessing of animal cruelty before the age of 13 was associated with higher perpetration rates (32%), compared to the witnessing of animal cruelty at age 13 or later (11.5%).

The above studies point to the importance of witnessing animal cruelty (i.e., an aggressive behaviour) for the learning of, and engagement in, aggressive behaviour. Children who witness or directly experience violence or aggression have been documented to be more likely to develop beliefs and scripts that support aggression (Guerra, Huesmann, & Spindler, 2003) and a tendency to behave aggressively (Anderson & Huesmann, 2003).

Much of the research examining the witnessing of cruelty has also examined engagement in cruel and violent behaviours. There is a great deal of overlap between the two, since youth are much more likely to witness cruelty and violence generally if they are raised in a violent family. Thompson and Gullone (2006) demonstrated that children who witnessed significant others engaging in animal cruelty were more likely to be cruel to animals themselves, compared to those who witnessed strangers engaging in animal cruelty. Similar findings were reported by Hensley and Tallichet (2005). They not only found that inmates who reported witnessing animal cruelty were more likely to be cruel to animals frequently but also that those who witnessed a family member or a friend hurting or killing animals were more likely to commit animal cruelty with greater frequency. Consistent with Henry's (2004) findings, and of developmental relevance, those who were younger when they first witnessed someone hurting or killing animals were more likely to commit animal cruelty more frequently.

Witnessing significant others such as parents abusing animals is likely to play a larger role in attitude formation for the child, contributing to the development of beliefs that aggressive and violent behaviours are somewhat normative, thereby supporting the development of what has been, in the general aggression literature, referred to as "normative beliefs" (Anderson & Huesmann, 2003). As previously noted, children's beliefs about aggression are correlated with those of their parents (Huesmann, Eton, Lefkowitz, & Walder, 1984) as well as those of their peers (Huesmann & Guerra, 1997).

Of course, it is not only the witnessing of aggression and violence that contributes to the learning of behaviour and to the formation of attitudes and beliefs; the actual experiencing of behaviour is likely to contribute to learning and attitude formation even more powerfully. The experiencing of abuse and neglect also interferes with the development of functional emotion regulation (Loeber & Hay, 1997) and the normative development of empathy (Zahn-Waxler, Radke-Yarrow, & King, 1979). Therefore, it is not at all surprising that a relationship has been found between children's experiences of abuse and neglect and their engagement in animal cruelty. The next section will review the research looking at the relationships between family and parenting experiences and children's animal cruelty.

Family and parenting experiences

In the earliest published investigation of the aetiology of animal cruelty by children, Tapia (1971) reported an analysis of 18 child cases of cruelty to animals selected from the clinic files of the Child Psychiatry Section of the University of Missouri's School of Medicine. In all selected cases, cruelty to animals was either the chief complaint or one of the referring complaints. Among the cases, there was a high male prevalence. The children were of normal intelligence and young in age, spanning from 5 to 15 years, with half of the cases being between 8 and 10 years. A chaotic home environment with aggressive parental models was the most common factor across the cases. On the basis of the case analysis, Tapia concluded that cruelty to animals occurs in conjunction with other hostile behaviour, including bullying and fighting, lying, stealing, and destructiveness, and that a chaotic home environment, together with aggressive parental models, are common factors.

A follow-up study was conducted in 1977 by Rigdon and Tapia to determine whether the presence of cruelty to animals as a significant clinical feature provides information that is of prognostic value. The original data reported in 1971 were collected between two and nine years earlier. Five of the original 18 children were not able to be located for this follow-up study. The detailed case-by-case analysis revealed that of the 13 cases followed up, eight were still cruel to animals as many as nine years later. The authors concluded that "Most of these children are the products of a chaotic home situation with aggressive parents who administered harsh corporal punishment" and that "The most effective form of therapy seemed to be removal from or a significant change in the chaotic home environment" (p. 36).

In other research, Deviney, Dickert, & Lockwood (1983) studied 53 families who had companion animals in their home and who met New Jersey legal criteria for child abuse and neglect. They found that compared to the general population, there were higher rates of animal cruelty in families where there was substantiated child abuse or neglect. Observations during home interviews revealed that companion animals were abused or neglected in 60% of these families. When the sample was classified according to type of abuse (physical abuse – 40%; sexual abuse – 10%; neglect – 58%), for an alarming 88% of families displaying physical abuse, cruelty to animals was also present. Two-thirds of the companion animals in these homes were abused by the fathers in the family, and one-third by children.

In their work comparing criminal (aggressive versus non-aggressive) and non-criminal retrospective reports of childhood experiences and abuse behaviours, Kellert and Felthous found that domestic violence, and particularly paternal abuse and alcoholism, were common factors among those aggressive criminals who had a history of childhood animal cruelty

(Felthous, 1980; Felthous & Kellert, 1986; Kellert & Felthous, 1985). According to Kellert and Felthous (1985), the family and childhood experiences of many of the aggressive criminals were particularly violent. The domestic violence in the families of the aggressive criminals was most strongly characterized by paternal violence. Of note, three-quarters of the aggressive criminals reported repeated and excessive child abuse, compared to 31% of the non-aggressive criminals and 10% of the non-criminals. Among the non-aggressive criminals and non-criminals who were cruel to animals, reports of being physically abused as children were common. As many as 75% of non-criminals who reported experiences of parental abuse also reported being cruel to animals.

In a study by Ressler, Burgess, Hartman, Douglas, & McCormack (1986), 36 convicted sexually-oriented killers were interviewed about their the childhood histories. The offenders who had been sexually abused in childhood or adolescence were significantly more likely than those who were not abused to report a number of aggressive behaviours, including cruelty to animals, cruelty to other children, and assaultive behaviour toward adults.

In research examining the relationships between childhood experiences and animal cruelty, Miller and Knutson (1997) compared the self-reports of 314 inmates in a corrections department with those of a group of undergraduate university students. They found modest associations between animal cruelty and punitive and acrimonious childhood histories. On this basis, the authors concluded that there is an association between punitive childhood histories and antisocial behaviour.

Also based on retrospective self-reports, Flynn's (1999b) study involved 267 undergraduate students. The results showed a relationship between corporal punishment by parents and the perpetration of animal cruelty. Those who had perpetrated animal cruelty were physically punished more frequently before the teenage years than those who had never been cruel to an animal. More than half of male teenagers who were hit by their fathers reported perpetrating animal cruelty.

Ascione, Friedrich, Heath, & Hayashi (2003) also examined the associations between children's cruelty to animals and physical abuse. In addition, they looked at the relationship between animal cruelty and parental physical fighting. Three groups of children (sexually abused group; psychiatric sample with no sexual abuse; and control group)aged between 6 and 12 years were involved in the study. Cruelty to animals was associated with a history of abuse, and the association was stronger for children who had been physically abused and those who had witnessed domestic violence.

A more recent study by Duncan, Thomas, & Miller (2005) yielded converging findings through the assessment of charts of boys (aged 8 to 17 years) with conduct problems. The children's histories were also examined to identify the occurrence of physical child abuse, sexual child abuse, paternal alcoholism, paternal unavailability, and domestic violence. Children were grouped according to whether they had or had not been cruel to animals. It was found that children who had been cruel to animals were two times as likely to have been physically and/or sexually abused or to have been exposed to domestic violence, compared to children who had not been cruel to animals.

In sum, these findings of research examining the relationships between childhood animal cruelty and parenting and family experiences are consistent with those from the larger literature relating to the development of antisocial behaviour. Such research, for example, has shown that within homes where there is greater family instability, more conflict, and problematic parenting strategies (e.g., physical punishment), children are more likely to development along the trajectory of childhood-onset antisocial behaviour, also noted as being the more problematic trajectory with regard to stability of aggression and severity of aggression.

Despite the methodological limitations that characterize many of the studies in the animal cruelty area, across different assessment methodologies, including retrospective reporting, a significant relationship between the experiencing of abuse in childhood and engagement in animal cruelty has emerged. Other factors placing children at risk of developing aggressive and antisocial behaviours, including animal cruelty behaviours, are those that characterize *risky families* (Repetti et al., 2002). These include overt family conflict, expressions of negative affect, and low nurturance and warmth. Risky parents are cold, unsupportive, or neglectful.

Risky parenting and risky family environments leave children vulnerable to the development of psychological and physical disorders. It is important to emphasize the interactional role played by both environment and biology. Whilst certain biologically-based characteristics, such as temperament, are predictive of developmental along an antisocial behaviour trajectory, children whose aggression increases as they develop, rather than following the normative decreasing pathway, may also be expressing a learnt survival behaviour for their particular circumstance. As victims of abuse, children experience a sense of powerlessness that, at a very basic level is likely to be experienced as a threat to survival. Identifying with their abuser enables a transformation from a sense of powerlessness to one of being in control. For a child, those who are more vulnerable than oneself are likely to be small animals, thus those animals are the vulnerable others onto whom aggression can be displaced.

Displacement of aggression

Displaced aggression, which has been described as a robust phenomenon (Marcus-Newhall, Pederson, Carlson, & Miller, 2000; Pederson, Gonzales, & Miller, 2000), constitutes a form of aggression against others (human or non-human animal) who did not play a direct role in the precipitating event. Displaced aggression increases if the target of such aggression provides even a minor trigger or the slightest of provocations (e.g., a dog barking). Displaced aggression also increases if the target can be perceived to be a member of a disliked out-group (Anderson & Huesmann, 2003) or as having less social value (e.g., a non-human animal).

There are instances when animal cruelty by children constitutes the displacement of aggression from humans to animals that occurs through the child's identification with their abuser. As previously noted, displaced aggression has been included as one of the nine motivations for animal cruelty reported by Kellert and Felthous (1985). By identifying with their abuser, a child's sense of powerlessness can be transformed into a sense of control or empowerment. Such an explanation is consistent with several hypothesized pathways of acquisition put forth within the human aggression literature.

As reviewed in Chapter 5, these include Bandura's (1983) social learning theory of aggression, which includes the proposal that the observation of aggressive models within the family, a sub-culture, or even the mass media can promote the development of aggressive behaviours. Moreover, as proposed by Huesmann (1988), aggressive scripts develop on the basis of attention to aggressive models. Such proposals shed light on why children growing up in a violent family, or children who are exposed to violent behaviour are more likely to behave aggressively, including engaging in animal cruelty. This line of thinking is also consistent with the research that has shown that parenting that is harsh or inconsistent, or parenting that is permissive, is predictive of child or youth aggression or antisocial behaviour (Burt, McGue, Krueger, & Iacono, 2005; Larsson, Viding, Rijsdijk, & Plomin, 2008; Loeber & Dishion, 1983; O'Connor, 2002; Repetti et al., 2002).

In addition to theories that focus upon the role of the family environment and of interpersonal relationships as risk factors for the development of animal cruelty behaviours, cognitive constructs such as those put forth within Information-Processing Models are relevant.

Cognitive errors, aggressive cues, and exposure to violence

Cognitive theories and constructs such as "knowledge structures" and perceptual schemas are useful in explaining the processes of acquisition of animal cruelty behaviours. Given that these structures are proposed to develop largely as a consequence of learning experiences, on the basis of theory it would be expected that individuals who experience or observe abuse in their formative years learn specific aggressive behaviours and hostile perceptions, attributions, and expectation biases. They are also more likely to learn callous attitudes and process to enable disengagement from normative empathic reactions that would otherwise serve as aggression inhibitors. The development of normative beliefs relating to antisocial behaviours is promoted by environments that are sympathetic toward such behaviours. Thus, for children with the relevant temperamental tendencies, "risky" families serve as appropriate incubators of Callous-Unemotional trait development. Experiences of abuse or neglect in childhood such as would occur within risky families, interfere with the otherwise normative development of empathy (Anderson & Bushman, 2002; Repetti et al., 2002). See Chapter 7 for a lengthier discussion of risky family environments.

Given the processes involved in the development of neural pathways as a consequence of our environmental experiences, we come to process information in particular ways, as has been detailed in Chapter 8. Compared to non-aggressive individuals, aggressive individuals are more likely to perceive hostility in situations even where there is none. This tendency, referred to as a "Hostile Attribution Bias", is particularly pronounced in ambiguous situations (Anderson & Bushman, 2002; Crick & Dodge, 1994; Dodge et al., 2006). In relation to animal cruelty, aggressive children may be more likely to attribute hostile intentions to animals, since cues provided by animals are often even more ambiguous than those provided by humans (Dadds, 2008). Such misattribution can go some way to explaining children's and even adults' aggression toward animals.

With regard to cognitive biases, as reviewed above, research has shown that it is not only those children who directly experience abuse who develop beliefs and scripts that support aggression or the tendency to behave violently, but also children who observe or witness abuse.

Whilst research has shown that witnessing significant others behaving in an aggressive manner serves as a powerful pathway of acquisition, observing media violence also has a significant effect on attitudes and behaviours (Anderson & Huesmann, 2003). As reviewed in Chapter 7, a large and robust body of research has consistently shown that exposure to media violence predicts an increase in aggressive thoughts, desensitization to later violence exposure, and reductions in physiological arousal following violence exposure. It also predicts an increased acceptance and endorsement of violent behaviour (Anderson & Huesmann, 2003; Anderson et al., 2010; Greeson & Williams, 1986; Hansen & Hansen, 1990).

There is therefore strong empirical evidence indicating that exposure to real-life or media violence plays a strong role in the formation of cognitions related to aggression and violence (Flynn, 1999b) as well as the development of aggressive behaviour (e.g. Baldry, 2005; Becker, et al., 2004; Currie, 2006; Gullone & Roberston, 2008; Margolin & Gordis, 2000; Thompson & Gullone, 2006).

These findings have significant implications for the role that legalized aggressive behaviours such as hunting, rodeos, and fishing have on the development of aggression, particularly for individuals with a vulnerable disposition toward the development of such behaviours, or those within a vulnerable environment or "risky" family. Labelling certain aggressive behaviours as entertainment or sport because they are targeting certain species, and others as antisocial because they are targeting other species such as companion animals or humans, is incongruous. Essentially, cruelty is legalized in instances that provide benefit to humans, such as confined farming practices for pork production. In contrast, for other species such as companion animal species, the behaviours that cause the suffering are outlawed. The potential psychological distress caused by such conflicting messages is managed, or effectively prevented from arising, through the use of cognitive mechanisms such as those referred to by Bandura in his Moral Disengagement Theory. As detailed in Chapter 8, these include, for example, vilifying the recipients, obscuring personal agency, or reconstructing the conduct. Such cognitive mechanisms enable us to disengage self-sanctions for engaging in reprehensible behaviour. For young people, whose attitudes are undergoing processes of formation, such contradiction and inconsistency can only serve as barriers to the development of empathy and compassion. Thus, these contradictory messages or differing standards of behaviour toward animals, depending upon the animal species and/or the context, themselves constitute risk factors for the development of aggression and violence.

Empathy and emotion regulation

As noted in the discussion of the normative development of aggression in Chapter 4, from as early as one year of age, aggression, particularly peer-directed aggression, becomes evident. By the time children have begun school, their aggression levels begin to decrease. It has been argued that this decrease coincides with an increase in interpersonal skills and emotion regulation competencies, including effortful control (Anderson & Huesmann, 2003; Eisenberg et al., 2004; Keenan & Shaw, 1997). Other developing abilities at this time include perspective-taking (Selman, 1980), empathy, (Zahn-Waxler, Radke-Yarrow, & King, 1979), and emotion processing (Schultz, Izard, & Bear, 2004). According to Ascione, Thompson, & Black (1997), motivations driving young children's animal cruelty, including *curiosity* and *exploration,* likely occur as a consequence of younger children not yet having internalized society's values regarding the appropriate treatment of animals. I would add that society's values regarding the appropriate treatment of animals are complex and confusing and do not lend themselves to being easily internalized until later years.

Consistent with a developmental psychopathology perspective, whilst the development of empathy and emotion regulation competencies predicts a decrease in aggressive behaviours, the compromised development of these abilities places children at risk of developing antisocial behaviours, including engaging in animal cruelty. Moreover, those children most at risk are likely to be that sub-group of children with Conduct Disorders who also present with Callous-Unemotional traits and an inability to experience guilt (Hastings, Zhan-Waxler, Robinson, Usher, & Bridges, 2000; Luk et al., 1999). These children tend to initiate and engage in persistent antisocial acts, including displays of aggression toward people and animals (Ascione, 1993; Miller, 2001). At this extreme end of the antisocial behaviour continuum, a lack of empathy and guilt, in addition to an interpersonal style characterized by callousness, are predictive of Psychopathy (Frick & White, 2008).

Whilst low levels of empathy constitute a risk factor for antisocial and aggressive behaviour (McPhedran, 2009), higher levels of empathy can be a protective factor against the development of these behaviours. Empathic and prosocial youths are more inclined to treat their companion animals humanely (Poresky 1990; Vidovic, Stetic and Bratko 1999). Several empirical studies have demonstrated the importance that empathy has for interpersonal relationships and behaviours, including those with animals. For example, Poresky's (1990) study assessed the relationship between bonds with companion animals and empathy levels among 38 children ranging in age from three to six. As expected, children who had a strong bond with their companion animal scored higher on empathy than those who did not have a companion animal.

In a related study, Vidovic, Stetic and Bratko (1999) assessed companion animal ownership and socio-emotional development among a sample of 826 youths ranging in age from 10 to 15. Participants who scored higher than average on a companion animal attachment scale yielded significantly higher scores on both empathy and prosocial orientation than those who scored lower than average. A more recent study involving 381 13- to 18-year-olds by Thompson and Gullone (2008) yielded supporting findings. These researchers examined the associations between empathy and prosocial behaviour as well empathy and antisocial behaviours. Behaviours toward humans and animals were investigated. As predicted, low empathy was found to be a significant predictor of antisocial behaviours, and high empathy was found to be a significant predictor of prosocial behaviours towards humans and animals.

Chapter summary

This chapter reviews and evaluates available research conducted within each of the two theoretical frameworks first reviewed in Chapter 5, beginning with research related to the Violence Graduation Hypothesis. The criticisms directed at this research have, in large part, caused the validity of this theory to be questioned. However, the review above highlights that the findings of the research conducted within this framework in fact provide empirical support for much of the general aggression research, particularly supporting conclusions based on individuals classified at the more severe end of the antisocial spectrum, including those who display Callous-Unemotional traits. The argument that violence graduates with age from less to more severe for people who show aggressive behaviours early in development is consistent with research looking at the childhood-onset group of aggressive individuals. Those individuals with childhood-onset aggression who also display Callous-Unemotional traits are most likely to display a life-course trajectory and to engage in behaviours characteristic of the more severe end of the antisocial spectrum.

Thus, the argument that these individuals graduate from animal abuse in childhood to human violence in adulthood is consistent with the findings related to the behaviours of individuals at this more severe end of the antisocial spectrum. Based on the general aggression literature and the research specifically examining the Violence Graduation Hypothesis, it is reasonable to propose that a pattern of repeated animal cruelty in young children is one behavioural marker of the developmental trajectory of life-course persistent and escalating aggression.

Research that has investigated the second dominant theoretical framework (that is, the Deviance Generalization Hypothesis) follows. Research noting the diagnostic importance of animal cruelty for Conduct Disorder and for Antisocial Personality Disorder is reviewed. This is followed by research examining the co-occurrence between animal cruelty and other criminal behaviours as well as other antisocial behaviours. Included is research that has demonstrated relationships between animal cruelty and bullying, juvenile firesetting, and criminal behaviours, including family or domestic violence. FBI work supports the importance of animal cruelty as a potential marker of particularly violent behaviour, such as serial killing and mass murder.

The final part of the chapter reviews research into animal cruelty that has investigated potential aetiological or risk factors for the development of animal cruelty behaviours. Understandings yielded from the general antisocial behaviour literature reviewed in Chapters 6, 7, and 8 are integrated with findings from animal cruelty research. Risk factors that are reviewed include the witnessing of violence, non-optimal family and parenting experiences (including child abuse and neglect), cognitive factors such as cognitive biases and errors, viewing of media violence, and emotional factors including emotion regulation.

When the findings of the animal cruelty literature are integrated with those from the extensive antisocial and aggressive behaviour literature, the inescapable conclusion is that repeated episodes of animal cruelty are a marker of the antisocial individual. Whilst more research can always refine understanding and advance knowledge, such research needs to be accurately positioned and comprehensively informed. It is only when armed with accurate conceptualization that animal cruelty researchers will contribute further and usefully to current understandings It is strongly recommended, therefore, that such conceptualization integrate understanding that has already been acquired from the broader aggression and antisocial behaviour literature.

In the next chapter, areas of investigation that are in need of further attention are highlighted. However, of more importance is the need for action based on currently existing knowledge. Most needed actions are those that will drive a change in attitudes and perceptions related to the suffering of animals. These issues are discussed in Chapter 10.

10
Conclusions and Future Directions

This Chapter concludes the book. It begins by examining two areas related to animal cruelty that have received little or no research attention. These are (i) the normative developmental patterns of animal cruelty, and (ii) cross-cultural comparisons of animal cruelty as defined herein. Following this section, focus turns to application of the vast research knowledge that supports a connection or link between different types of abuse and cruelty across humans and animals. In particular, scholars have argued that it is now time for action on the part of law and policy makers. Such action is needed to punish animal cruelty behaviour as well as to deter future animal cruelty behaviour. Given the strong and mutual relationship between the law and public perceptions, such action will, over time, bring about a change in public perception about the importance of animal cruelty, thereby heightening societal empathy expectations and consequently lowering tolerance levels for antisocial behaviour.

The development of animal cruelty behaviour

There has been a significant amount of research attention devoted to mapping out the developmental pattern of antisocial and aggressive behaviour generally (see Chapter 4), however, there is a paucity of research into the development of animal cruelty behaviour. Whilst research has provided understanding of the risk factors for the development of animal cruelty behaviours such as experiences of child abuse and neglect (e.g., Alys et al., 2009; Felthous & Kellert, 1987; Merz-Perz et al., 2001), there is no existing study that has specifically examined the normative development of animal cruelty. Some have commented on the higher levels of animal cruelty in early childhood. MacDonald

(1979) argued that such behaviour occurs in many children as a developmental stage. This view was echoed by Ascione and colleagues some decades later (Ascione & Lockwood, 2001; Ascione, Thompson & Black, 1997) and is consistent with the developmental literature on aggression.

In the general aggression literature, the normative pattern of a decrease in aggressive behaviour in the early to middle childhood years has been noted. This implies a higher level of aggression prior to the decrease which is argued to result from increasing interpersonal skills. It is argued that children are socialized out of their earlier aggressive behaviours. It is also argued that there is a relationship with children's development of their ability to control or regulate their emotions, otherwise referred to as "effortful control" (Eisenberg, et al., 2004; Keenan & Shaw, 2003). Empathy also increases during this period (Zahn-Waxler, et al., 1979), as does emotion processing (Schultz, et al., 2004).

Thus, as with many other facets of animal cruelty research, there are consistencies with the general aggression and antisocial behaviour literature. Nevertheless, at present, the views put forth about the normative development of animal cruelty behaviour are theoretical. They have not been empirically supported with research. It is hypothesized that investigation of the normative development of animal cruelty would mirror the normative development of aggressive behaviour. Supportive findings would serve to reinforce the argument that animal cruelty behaviours are accurately conceptualized within the antisocial behaviour framework.

Cross-cultural research

Another neglected area is investigation of the role played by culture in animal cruelty behaviours. Given the vast differences in the treatment of animals across cultures, it is reasonable to assume that animal cruelty behaviours are conceptualized and perceived differently across culture. For example, in many Asian countries, the eating of animals that are given companion animal status in Western countries is commonplace. Laws governing animal welfare are also variable and in many cases not well monitored. For example, bear bile farms are illegal in Vietnam. Despite this, many bears continue to be farmed for their bile. The slaughter of animals is also managed to varying degrees such that the legal requirement of prior stunning is absent in many countries and cultures (e.g., Indonesia).

Given these differences, it is logical to hypothesize that the conceptualization of animal cruelty as deviant, as can be argued for Western nations and cultures, will have varying validity, depending on that culture's animal treatment standards. Although further research is needed, the few studies that have been conducted in cultures other than Australia, the United States, and the United Kingdom shed some light on this issue.

A study by Pagani, Robustelli, and Ascione (2007) examined the attitudes on Italian youth and their concern for animals. Although of interest given that this study researched a population other than the more commonly researched American culture, the Italian culture is documented in the cross-cultural psychology literature to be a Western culture (Matsumoto, & Juang, 2008). Of more relevance to the current discussion is the study conducted by Mellor, Yeow, Mamat, and Hapidzal (2008) which investigated attitudes toward animals in Malaysian culture, and that conducted by Mellor and colleagues (2009), which investigated children's animal cruelty across three cultures (Japan, Malaysia, Australia). These studies will be reviewed in more detail following a brief summary of the Pagani et al., (2007) study.

Pagani et al. (2007) investigated different aspects of child-animal relationships in the Italian culture. This included attitudes toward companion animal ownership, animal cruelty as both a perpetrator and a witness, and animals as sources of fear and comfort. Also examined were attitudes toward hunting and the use of fur and leather clothing. Twelve schools were involved in the research and ranged from primary to high schools, and a sample ranging in age from 9 to 18 years was recruited. A total of 800 youth equally representing male and female children and adolescents were involved in the study.

The findings revealed that children who had a secure or healthy attachment to their companion animals were more likely to be opposed to hunting and zoos. They were also more likely to express concern over roadkill. Regarding animal cruelty, a low, although notable, frequency was found. Sex differences were also pronounced with boys admitting to more acts of animal cruelty compared to girls. Common motives reported by the youth for engaging in animal cruelty included: fun or pleasure, aversion or indifference toward animals, or a method of discipline or punishment. The authors reported that fear of animals as well as low empathy levels predicted more animal cruelty. Additionally, and as one would expect, a healthy attachment to companion animals was predictive of less animal cruelty. Finally, children's exposure to animal cruelty was documented as occurring with a very high frequency.

This study yielded findings consistent with much of the other research that has examined children's animal cruelty with the exception of the very high frequency of animal cruelty reported by the young sample. These findings are at odds with much of the research that has investigated self-reported cruelty. Whilst it may be a culturally relevant finding, it would need to be replicated before any conclusion can be drawn. Given the consistency of the other findings with existing research, the differences in outcomes are more likely due to methodological or sample specifics. Thus, unfortunately this study does not make a significant contribution to the role that culture plays in animal cruelty.

The study conducted by Mellor and colleagues (2008) provides an important addition to the human-animal relationships literature. Their study involved 379 Malaysian children aged 6 to 12 years. (See Al-Kaysi, 1986; and Banderker, undated for descriptions of the ways animals are viewed in this culture.) They examined relationships between parent-reported animal cruelty and parent- as well as child-reported psychological strengths and weaknesses. As has been found in other research (e.g., Ascione, 1993; Dadds, Whiting, & Hawes, 2006; and Frick et al., 1993), the results indicated that children's animal cruelty was associated with externalizing difficulties including Conduct Disorders and hyperactivity. This is an important finding, since it indicates that even when attitudes toward animals vary, animal cruelty remains a significant predictor of deviant or disordered behaviour.

Mellor and colleagues (2009) conducted possibly the only existing cross-cultural study examining childhood cruelty to animals. Their sample included 1,358 children aged between 5 and 13 years from Japan (313 boys, 321 girls), Australia (185 boys, 163 girls), and Malaysia (149 boys, 227 girls). Prior to describing their study, the authors provided an overview of the ways in which animals are viewed and the animal welfare standards in the three countries.

Mellor et al.'s paper describes Japan as a country which has historically not given a great deal of consideration to animal welfare despite its economically developed and technologically advanced status. Nevertheless, there is a Japanese Society for the Prevention of Cruelty to Animals, and there is legislation relating to animal welfare. However, Japanese law has no definition of animal cruelty, and the country's animal protection law has rarely been invoked. Moreover, there is no government department that oversees animal welfare. Companion animal ownership is popular in Japan, with around 48.4% of homes having companion animals. Dogs, cats, and fish are the most common types of animals. The popularity of companion animals in Japan, however, reflects the culture's fashion consciousness rather than a love of animals (Karasaki, 1999). Different breeds of dogs become fashionable at different times, and the discarding of dogs when that particular breed is no longer fashionable is reflected in statistics related to abandonment of animals. Mellor and colleagues note that in 1997, around 400,000 dogs and 300,000 cats were destroyed in Japan. Japan is also a market for exotic companion animals, with a variety of species being purchased as companions, including ponies, pigs, wallabies, owls, eagles, and monkeys.

Malaysia's interactions between humans and animals are described as being largely influenced by religion. Some cultures, such as the Islamic culture, see dogs as unhygienic, and it is therefore not permissible, according to the religion, to keep dogs in the home. It is also forbidden to touch dogs. However, it is necessary for people who own animals to provide them with adequate shelter, food, and water. It is also necessary to provide the animal with veterinary attention when necessary. In Malaysia, there is a society for the prevention of cruelty to animals (Al-Kaysi, 1986; Banderker, undated).

Australia is a country which has a multicultural population that is predominantly Christian. It has one of the highest rates of companion animal ownership in the world, with around 63% of households in 2005 owning at least one companion animal (Australian Companion Animal Council, 2007). Each state and territory has its own animal welfare acts, and the federal government makes some attempt to ensure that the legislation is uniform across the different jurisdictions. Among the many animal welfare organizations is the Royal Society for the Prevention of Cruelty Against Animals.

The parents of the participants were asked to complete a questionnaire asking about their children's animal cruelty (Guymer, Mellor, Luk, & Pearse, 2001). The results showed that, on the whole, boys were more likely than girls to exhibit cruelty. However, sex differences were not present for adolescents aged between 11 and 13 years. It was also found that younger Australian and Japanese children, but not Malaysian children, were more cruel to animals than were older children.

One significant limitation of the study is that the findings were based on parent reports, and it has been documented that parents underestimate their children's cruelty. The finding that younger children were more cruel than older children may be an artefact of the method, since parents are less likely to be aware of their adolescent children's cruelty than they are of their younger children's cruelty. Finally, the lack of differences across the three countries in children's cruelty, despite quite notable differences in the cultures, suggests universalities in the levels of children's animal cruelty. However, the lack of difference may also reflect insensitivity of the measures used in the different cultures. The application of Western measures to non-Western cultures has been referred to as "imposed etic" and has been identified as a significant limitation of many cross-cultural studies (Matsumoto, & Juang, 2008). Given the very limited cross-cultural research that has been conducted in both the animal cruelty area and in the antisocial behaviour area, more research is called for.

An action agenda

Of importance, but not covered in this work, are the implications of research on animal cruelty for psychological interventions aimed at prevention and treatment. The broader aggression and antisocial aggression literature includes a great deal of work examining preventative and treatment strategies, and the reader is referred to a number of sources for more detailed information (e.g., Barlow, 2006; Cicchetti & Toth, 2006; Faver, 2010; Fraser et al., 2005; Walter, 2001; Worsham & Goodvin, 2007; Zigler, Taussig, & Black, 1992). For the reasons that follow, however, reviewing this work here is considered to be outside the scope of this work.

It is a basic argument of my thesis that the core problematic issue relating to the best way to address animal cruelty does not stem from a lack of prevention and intervention knowledge. Rather, the problem results from a lack of perceived worthiness of animal cruelty as a target for intervention. A change is needed in the perceived importance of animal cruelty among researchers, health care professionals, and legislators. Such a change may slowly come about if animal cruelty becomes more strongly recognized and accepted as a characteristic behaviour of Antisocial individuals or as a symptom of Antisocial Disorders such as Conduct Disorder or Psychopathy.

In an ideal world, animal cruelty would not need to be linked with human harm to be considered worthy of moral consideration. Unfortunately, however, as history has shown with regard to the relative neglect of animals in the broader scientific community, clear evidence for the strong link between animal cruelty and behaviours that harm humans is needed to change people's values and perceptions about the importance of animal cruelty as a valid marker of the antisocial or aggressive individual.

Proposed strategies for change

As argued in Chapter 1, the general position held by society and its members is that animals' suffering, when compared to humans', is less worthy of both scientific and moral consideration. Not surprisingly, this perception has resulted in animal cruelty receiving less scientific attention. The flow-on effect of this perception has been the rather slow recognition of the important link that is evident across different abusive behaviours not only by the scientific community but the legal system as well. Since public opinion is related to the law of the land, and likewise the law of the land reflects public opinion, this attitude toward animal suffering has become stuck in a vicious cycle of discrimination. The purpose of this book has been to highlight the strong empirical evidence that links different types of abuse. It is now time for action.

A number of authors (e.g., Clawson, 2009; Lockwood, 2008; Schaffner, 2009) have put forth possible strategies for the promotion of the cultural change needed to enable animal cruelty to be appropriately regarded as the antisocial and aggressive behaviour that it is. These strategies include increasing public awareness of animal cruelty and related issues, as well as strengthening animal cruelty legislation. Whilst law often reflects public perceptions, it also reflects scientific knowledge and consequently brings about a change in public perceptions. In this instance, as has happened with child and partner abuse, changes in the law and legal proceedings are necessary to bring about changes in public perceptions.

First, at the most basic level, the status of animals needs to change in the eyes of the law so that animals are perceived as more than mere property. Additionally, legislature needs to enact cruelty laws that appropriately reflect the severity of the offence (Clawson, 2009; Lockwood, 2008; Schaffner, 2009). As argued by Schaffner (2009), "the law should punish violent criminals according to the acts that they perpetrate. Whether the victim is a human being or an animal, a violent crime is a crime against its intended victim, as well a crime against society and its morals" (p. 199).

It goes without saying that the general perception of animal cruelty as relatively unimportant when compared to crimes against humans, is having a negative impact on animals. However, as highlighted throughout much of this work, the same perception is also having a negative impact on humans, since many crimes against humans may well have been prevented had any animal cruelty incidents that preceded them been taken more seriously. As cogently stated by Schleuter (2008):

> Most agencies consider crimes against animals complaints to be low-priority calls, regardless of whether they are in-progress crimes, and no matter how violent. ... If one pays attention to such crimes occurring particularly in a dysfunctional intra-familial setting or in conjunction with other destructive behaviours, future animal abuse or neglect, as well as similar crimes against vulnerable members of the household, may become potentially preventable if proper interventions are put in place. (p. 378)

Related to acknowledgement of the link between animal cruelty and human-directed aggression or violence is a particularly important strategy that several authors have proposed. The strategy is the facilitation of cross-reporting of suspected animal cruelty. Such cross-reporting would involve reporting suspected animal cruelty not only to animal welfare organizations but also to the police force and human service agencies such as child protective services and adult protective services.

In his chapter specifically addressing the role of laws and policy to address the link of family violence, Schaffner (2009) acknowledges that although social scientists have provided data to support the link, the law has been slow to respond. Current laws independently address crimes, depending upon the victim. Thus, there are separate laws for animal cruelty, family or domestic violence, child abuse, and elder abuse. There is no existing law that directly targets the relationships among these different crimes of abuse.

In providing justification for the proposed need for a law that recognizes the link between different types of abuse, among other reasons, Schaffner argues that one act of abuse often involves multiple forms of violence such that the abuse of a companion animal also psychologically harms other members of the family. According to Schaffner, if properly implemented, law and policy would be better armed to: (i) prevent human abuse, (ii) detect abuse earlier than is presently the case, (iii) protect family violence victims by providing protective orders and safe havens, (iv) facilitate prosecution of the abuser or perpetrator, and (v) avoid further abuses through provision of appropriate sanctions. It is also true that a law directly targeting the relationships between different crimes of abuse would aid in prevention of animal cruelty.

Thus, by enacting adequate animal cruelty laws that properly indicate the seriousness of the animal cruelty crime committed, future violence toward both human and animal victims can be prevented. To properly reflect the seriousness of the offence, the law should impose a minimum sentence (Schaffner, 2009). If the severity of animal cruelty crimes is adequately signalled by law and sanctions, the perceptions of prosecutors will change so that they begin to reflect current scientific understanding in their decisions as opposed to falling prey to existing biases that animals are less worthy of moral consideration. It is ultimately the responsibility of prosecutors to enforce animal cruelty laws by prosecuting animal cruelty cases to the fullest extent of the law.

Concluding comments

The good news is that, despite the slowness of progress regarding appropriate recognition of animal cruelty, in recent years there has been an increasing acceptance of the link between human antisocial behaviour and animal cruelty. This is reflected in the increased attention that animal cruelty has received in the media (e.g., Urbina, 2010). It is also reflected in the changing laws related to animal cruelty. For example, in the United States, at least 27 states now allow courts to bar animal cruelty perpetrators from owning or coming into contact with companion animals if they have been convicted of a cruelty crime. In addition, more than 30 U.S. states now have laws that shift the financial burden for the caring of abused or neglected animals to those convicted. Thirty U.S. states also currently authorize the reporting of suspected animal cruelty by veterinarians. Further, reflecting acknowledgement of the link between animal cruelty and human violence, and heeding the call for cross-reporting, eight U.S. states now have laws that authorize child or spousal abuse investigators and animal control officers to inform each other when they suspect cruelty.

So it is that in the 21st century, a shift in perceptions seems to be slowly taking place such that we are finally beginning to witness a renaissance of the views expressed by philosophers and anthropologists centuries ago that strongly declared a link between violence against animals and violence against people. The remaining most significant challenge is to have these views enshrined in laws that are fairly and fully implemented.

Chapter summary

Two animal cruelty areas in need of research are (i) examination of the normative developmental patterns of animal cruelty, and (ii) the cross-cultural universality versus specificity of relationships between animal cruelty and human antisocial behaviour. Regarding the first area in need of research, there has been some argument that animal cruelty is normatively high in the very early childhood years. The high frequency of the behaviour early in development has been argued to be exploratory in nature and motivated by curiosity. Consistent with the normative literature on aggression, proposals have been made that early childhood animal cruelty behaviour decreases once children begin to internalize social norms and develop increased interpersonal skills. An increase in emotion-regulation strategies, such as effortful control, is another possible explanatory factor in the normative decrease of animal cruelty as children develop into the middle childhood years. However, to date there has been no empirical investigation of the normative developmental patterns of animal cruelty behaviour. Such investigation is needed.

The second area identified as being in need of empirical investigation is the cultural specificity or universality of animal cruelty behaviours and attitudes toward such behaviours. Given differences in attitudes toward animals across different cultures and nations, and given different standards of behaviour toward them, one important question relates to whether the same relationships between animal cruelty and human antisocial behaviour can be found across different cultures and nations. Findings of cross-cultural research will have implications for the generalizability across cultures of the main argument being made in the current work.

The final part of this chapter changes focus from research to practice. Certainly, in the West, where much of the aggression and antisocial behaviour research has been conducted, there is sufficient evidence that there are strong overlaps between animal cruelty and other abuses. Indeed, as demonstrated throughout this book, it can be argued confidently that animal cruelty is yet another marker of antisocial or aggressive behaviour that can be classified along the externalizing behaviour spectrum. Thus, the research outcomes are clear.

It is now time for law and policy makers to act upon this vast body of research. Laws need to be developed that acknowledge the relationship and similarities between different types of abuse and violence, including animal cruelty. Consistent with this recognition, laws need to be implemented fully and fairly across different abusive behaviours. Based on the empirical information available, there exists no possible justification for relegating animal cruelty offences to the "less important" category. Consequently, there is no justification for punishing violent criminals significantly more leniently or, as often happens, not at all, if the victim of their violent crime is an animal as opposed to a human being. Indeed, there is a high statistical probability that the victims of the violent or antisocial individual are both animal and human. This strengthens the argument that laws should punish criminals according to the severity of the acts they perpetrate, without discrimination or favour based on the target species of the particular crime.

References

Achenbach, T.M. (1966). The classification of children's psychiatric symptoms: a factor analytic study. *Psychological Monographs, 80 (1,* Whole No. 615).

Achenbach, T.M. (1978). The child behavior profile: I. boys aged 6–11. *Journal of Consulting and Clinical Psychology, 46,* 478–488.

Achenbach, T.M., & Edelbrock, C.S. (1979). Child behavior profile: II. boys aged 12–16 and girls aged 6–11 and 12–16. *Journal of Consulting and Clinical Psychology, 47,* 223–233.

Achenbach, T.M., & Edelbrock, C.S. (1983). *Manual for the Child Behavior Checklist and Revised Child Profile Behavior Profile.* Burlington: University of Vermont, Department of Psychiatry.

Affifi, T.O., Brownridge, D.A., Cox, B.J., & Jitender, S. (2006). Physical punishment, childhood abuse and psychiatric disorders. *Child Abuse & Neglect, 30,* 1093–1103.

Ainsworth, M.D.S., & Eichberg, C. (1991). Effects on infant-mother attachment of mother's unresolved loss of an attachment figure, or other traumatic experience. In C.M. Parkes, J. Stevenson-Hinde & P. Marris (eds.), *Attachment across the life cycle* (pp. 160–183). New York: Routledge.

Alessandri, S.M. (1991). Play and social behavior in maltreated preschoolers. *Developmental Psychopathology, 3,* 191–205.

Al-Kaysi, M.I. (1986). *Morals and Manners in Islam: a Guide to Islam Adab.* Leicester: Islamic Foundation.

Alys, L., Wilson, J.C., Clarke, J., & Toman, P. (2009). Developmental animal cruelty and its correlates in sexual homicide offenders and sex offenders. In A. Linzey (ed.), *The Link between Animal Abuse and Human Violence* (pp. 145–162). Eastbourne: Sussex Academic Press.

American Psychiatric Association. (1987). *Diagnostic and Statistical Manual of Mental Disorders* (3rd ed). Washington, D.C: Author.

American Psychiatric Association. (1994). *Diagnostic and Statistical Manual of Mental Disorders* (4th ed). Washington, D.C: Author.

American Psychiatric Association. (2000). *Diagnostic and Statistical Manual of Mental Disorders* (4th ed). Text Revised. Washington, D.C: Author.

Anderson, C. A. (2000). Violence and aggression. In A.E. Kazdin (ed.), *Encyclopedia of Psychology, Vol. 8.* (pp. 162–169). New York: Oxford University Press.

Anderson, C.A. (2002). Aggression. (p. 68–78). In E. Borgatta & R. Montgomery (eds.). *The Encyclopedia of Sociology* (2nd ed.), New York: MacMillan.

Anderson, C.A. Benjamin, A.J., & Bartholow, B.D. (1998). Does the gun pull the trigger? Automatic priming effects of weapon pictures and weapon names. *Psychological Science, 9,* 308–314.

Anderson, C.A., & Bushman, B. J. (2002). Human aggression. *Annual Review of Psychology, 53,* 27–51.

Anderson, C.A., & Dill, K.E. (2000). Video games and aggressive thoughts, feelings, and behavior in the laboratory and in life. *Journal of Personality and Social Psychology, 78,* 772–790.

Anderson, C.A. & Huesmann, L.R. (2003). Human aggression: A social-cognitive view. In M.A. Hogg & J. Cooper (eds.). *The Sage Handbook of Social Psychology.* (pp. 296–323). Thousand Oaks, CA: Sage Publications.

Anderson, C.A., Shibuya, A., Ihori, N., Swing, E. L., Bushman, B. J., Sakamoto, A., et al. (2010). Violent video game effects on aggression, empathy, and prosocial behavior in Eastern and Western countries: A meta-analytic review. *Psychological Bulletin, 136,* 151–173.

Arluke, A., Levin, J., Luke, C., & Ascione, F. (1999). The relationship of animal abuse to violence and other forms of antisocial behavior. *Journal of Interpersonal Violence, 14,* 963–975.

Arluke, A., & Luke, C. (1997). Physical cruelty toward animals in Massachusetts, 1975–1996. *Society and Animals, 5,* 195–204.

Ascione, F.R. (1993). Children who are cruel to animals: A review of research and implications for developmental psychopathology. *Anthrozoos, 6,* 226–247.

Ascione, F.R. (1998). Battered women's reports of their partners' and their children's cruelty to animals. *Journal of Emotional Abuse, 1,* 119–133.

Ascione, F.R. (1999). The abuse of animals and human interpersonal violence: Making the connection. In F. Ascione & P. Arkow (eds.), *Child Abuse, Domestic Violence, and Animal Abuse: Linking the Circles of Compassion for Prevention and Intervention* (pp. 50–61). West Lafayette, Indiana: Purdue University Press.

Ascione, F.R. (2001). Animal abuse and youth violence. *Juvenile Justice Bulletin,* September.

Ascione, F. (2005). Children, Animal Abuse, and Family Violence – The multiple intersections of animal abuse, child victimization and domestic violence. In K. A. Kendall-Tackett & S.M. Giacomoni (eds.), *Child Victimization: Maltreatment, Bullying and Dating Violence, Prevention and Intervention.* Kingston, NJ: Civic Research Institute.

Ascione, F.R., Friedrich, W.N., Heath, J., & Hayashi, K. (2003). Cruelty to animals in normative, sexually abused, and outpatient psychiatric samples of 6- to 12-year-old children: Relations to maltreatment and exposure to domestic violence. *Anthrozoos, 16,* 194–212.

Ascione, F.R., & Lockwood, R. (2001). Cruelty to animals: Changing psychological, social, and legislative perspectives. In Humane Society of the United States (ed.), *The state of animals.* (pp. 39–64). Gaithersburg, M.D., Humane Society Press.

Ascione, F.R., Thompson, T.M., & Black, T. (1997). Childhood cruelty to animals: Assessing cruelty dimensions and motivations. *Anthrozoos, 10,* 170–179.

Ascione, F.R., Weber, C. V., Thompson, T.M., Heath, J., Maruyama, M., & Hayashi, K. (2007). Battered pets and domestic violence: Animal abuse reported by women experiencing intimate violence and by nonabused women. *Violence Against Women, 13,* 354–373.

Australian Companion Animal Council (2007). Pet ownership statistics. Available at http://www.acac.org.au/. Retrieved 24 September 2010.

Baldry, A.C. (1998). Bullying among Italian middle school students: Combining methods to understand aggressive behaviours and victimisation. *School Psychology International, 19,* 361–374.

Baldry, A.C. (2003). Animal abuse and exposure to interparental violence in Italian youth. *Journal of Interpersonal Violence, 18*(3), 258–281.

Baldry, A.C. (2005). Animal abuse among preadolescents directly and indirectly victimized at school and at home. *Criminal Behaviour and Mental Health, 15,* 97–110.

Baldry, A.C., & Farrington, D.P. (2000). Bullies and delinquents: Personal characteristics and parental styles. *Journal of Community & Applied Social Psychology, 10,* 17–31.

Banderker, A.M. (undated). Dogs in Islam. Animal voice. Available at http://www.islamicconcern.com/dogs.asp Retrieved 12 November 2011.

Bandura, A. (1973). *Aggression: A Social Learning Theory Analysis.* Englewood Cliffs, NJ: Prentice-Hall.

Bandura, A. (1977). *Social Learning Theory.* Oxford, England: Prentice-Hall.

Bandura, A. (1978). Social learning theory of aggression. *Journal of Communication, Summer,* 12–29.

Bandura, A. (1983). Psychological mechanisms of aggression. In R.G. Geen & E.I. Donnerstein (eds.), *Aggression: Theoretical and Empirical Reviews, Vol. 1.* (pp. 1–40). New York: Academic Press.

Bandura, A. (1990). Selective activation and disengagement of moral control. *Journal of Social Issues, 46,* 27–46.

Bandura, A. (1999). Moral disengagement in the perpetration of inhumanities. *Personality and Social Psychology Review, 3,* 193–209.

Bandura, A., Barbaranelli, C., Caprara, G.V., & Pastorelli, C. (1996). Mechanisms of moral disengagement in the exercise of moral agency. *Journal of Personality and Social Psychology, 71,* 364–374.

Bandura, A., Underwood, B., & Fromson, M.E. (1975). Disinhibition of aggression through diffusion of responsibility and dehumanization of victims. *Journal of Research in Personality, 9,* 253–269.

Bandura, A., & Walters, R.H. (1959). *Adolescent Aggression.* New York: Ronald.

Bargh, J.A., Lombardi, W.J., & Higgins, E.T. (1988). Automaticity of chronically accessible constructs in person X situation effects on person perception: It's just a matter of time. *Journal of Personality and Social Psychology, 55,* 599–605.

Bargh, J.A., & Pietromonaco, P. (1982). Automatic information processing and social perception: The influence of trait information presented outside of conscious awareness on impression formation. *Journal of Personality and Social Psychology, 43,* 7–49.

Bariola, E., Gullone, E., & Hughes, E.K. (2011). Child and adolescent emotion regulation: The role of parental emotion regulation and expression. *Clinical Child and Family Psychology Review, 14,* 198–212.

Barlow, J. (2006). Individual and group-based parenting programmes for the treatment of physical child abuse and neglect. *Cochrane Database of Systematic Reviews,* CD005463.

Baron, R.A. (1999). Social and personal determinants of workplace aggression: Evidence for the impact of perceived injustice and the Type A behavior pattern. *Aggressive Behavior, 25,* 281–296.

Bates, J.E., Pettit, G.S., Dodge, K.A., & Ridge, B. (1998). Interaction of temperament resistance to control and restrictive parenting in the development of externalizing behavior. *Developmental Psychology, 34,* 982–995.

Baumeister, R.F., & Boden, J.M. (1998). Aggression and the self: High self-esteem, low self-esteem, and ego threat. In R.G. Geen & Donnerstein, E. (eds.), *Aggression: Theoretical and Empirical Reviews, Vol. 1.* (pp. 111–137), New York: Academic Press.

Becker, K.D., Stuewig, J., Herrera, V.M., & McCloskey, L.A. (2004). A study of firesetting and animal cruelty in children: Family influences and adolescent outcomes. *Journal of the American Academy of Child & Adolescent Psychiatry, 43,* 905–912.

Beetz, A.M. (2008). Bestiality and zoophilia: A discussion of sexual contact with animals. In F.R. Ascione (ed.). *The International Handbook of Animal Abuse and Cruelty: Theory, Research, and Application* (pp. 201–220). West Lafayette, Indiana: Purdue University Press.

Beirne, P. (2004). From animal abuse to interhuman violence? A critical review of the progression thesis. *Society & Animals, 12,* 39–65.

Berkowitz, L. (1962). *Aggression: a Social Psychological Analysis.* New York: McGraw Hill.

Berkowitz, L. (1989). Frustration-Aggression Hypothesis: Examination and reformulation. *Psychological Bulletin, 106,* 59–73.

Berkowitz, L. (1993). *Aggression: Its Causes, Consequences and Control.* New York: McGraw Hill.

Beyers, J.M., Bates, J.E., Pettit, G.S., Dodge, K.A. (2003). Neighborhood structure, parenting processes, and the development of youths' externalizing behaviors: A multilevel analysis. *American Journal of Community Psychology, 31,* 35–53.

Bierman, K.L., Smoot, D.L., Aumiller, K. (1993). Characteristics of Aggressive-Rejected, Aggressive (Nonrejected), and Rejected (Nonaggressive) Boys. *Child Development, 64,* 139–151.

Black, D.A., Heyman, R.E., & Smith Slep, A.M. (2001). Risk factors for child physical abuse. *Aggression and Violent Behavior* 6: 121–188.

Blake, C.S., & Hamrin, V. (2007). Current approaches to the assessment and management of anger and aggression in youth: A review. *Journal of Child and Adolescent Psychiatric Nursing, 20,* 209–221.

Blair, R.J.R. (1999). Responsiveness to distress cues in the child with psychopathic tendencies. *Personality and Individual Differences, 27,* 135–145.

Blair, R.J.R., Peschardt, K.S., Budhani, S., Mitchell, D.G.V., & Pine, D.S. (2006). The development of psychopathy. *Journal of Child Psychology and Psychiatry. 47,* 262–275.

Blonigen, D.D., Hicks, B.M., Kruger, R.F., Patrick, C.P., & Iacono, W.G. (2006). Continuity and change in psychopathic traits as measured via normal-range personality: A longitudinal-biometric study. *Journal of Abnormal Psychology, 115,* 85–95.

Blumstein, A. (2000). Disaggregating the violence trends. In A. Blumstein & J. Wallman (eds.), *The crime drop in America.* (pp. 13–44). New York: Cambridge University Press.

Blumstein, A., & Cohen, J. (1987). Characterizing criminal careers. *Science, 237,* 985–991.

Bosworth, K., Espelage, D.L., & Simon, T.R. (1999). Factors associated with bullying behavior in middle school students. *Journal of Early Adolescence, 19,* 341–362.

Bolinger, D. (1982). *Language: The Loaded Weapon.* London: Longman.

Booth, C.L., Rubin, K.H., & Rose-Krasnor, L. (1998). Perceptions of emotional support from mother and friend in middle childhood: Links with social-emotional adaptation and preschool attachment security. *Child Development. 69,* 427–442.

Bor, W., McGee, T.R., & Fagan, A.A. (2004). Early risk factors for adolescent antisocial behaviour: An Australian longitudinal study. *Australian and New Zealand Journal of Psychiatry, 38*, 365–372.

Bowd, A.D., & Bowd, A.C. (1989). Attitudes toward the treatment of animals: A study of Christian groups in Australia. *Anthrozoos, 3*, 20–24.

Bowlby, J. (1969). *Attachment and Loss, Vol. 1: Attachment.* New York: Basic Books (2nd rev. ed., 1982).

Brantley, A.C. (September, 2007). The use of animal cruelty evidence in dangerousness assessments by law enforcement. 1st International Academic Conference on the Relationship between Animal Abuse and Human Violence, Oxford Centre for Animal Ethics, Oxford, UK.

Brock, T.C., & Buss, A.H. (1962). Dissonance, aggression, and evaluation of pain. *Journal of Abnormal and Social Psychology, 65*, 197–202.

Broom, D.M. (1991). Animal welfare: Concepts and measurements. *Journal of Animal Science, 69*, 4167–4175.

Brown, L. (1988). *Cruelty to Animals: The Moral Debt.* London: Macmillan.

Bryant, B.K. (1990). The richness of the child-pet relationship: A consideration of both benefits and costs of pets to children. *Anthrozoos, 3*, 253–261.

Burke, J.D., Loeber, R., & Lahey, B.B. (2007). Adolescent conduct disorder and interpersonal callousness as predictors of psychopathy in young adults. *Journal of Clinical Child and Adolescent Psychology, 36*, 334–346.

Burt, S.A. (2012). How do we optimally conceptualize the heterogeneity within antisocial behavior? An argument for aggressive versus non-aggressive behavioral dimensions. *Clinical Psychology Review, 32*, 263–279.

Burt, S.A., McGue, M., Krueger, R.F., & Iacono. W.G. (2005). How are parent-child conflict and childhood externalizing symptoms related over time? Results from a genetically cross-lagged study. *Development and Psychopathology, 17*, 145–165.

Bushman, B.J. (1998). Priming effects of violent media on the accessibility of aggressive constructs in memory. *Personality and Social Psychology Bulletin, 24*, 537–545.

Bushman, B.J., & Anderson, C.A. (2001). Media violence and the American public. Scientific facts versus media misinformation. *American Psychologist, 56*, 477–489.

Carlisle-Frank, P., Frank, J.M., & Nielsen, L. (2004). Selective battering of the family pet. *Anthrozoos, 17*, 26–42.

Carlson, E.A., & Sroufe, L.A. (1995). Contribution of attachment theory to developmental psychopathology. In D. Cicchetti & D.J. Cohen (eds.), *Developmental Psychopathology: Theory and Methods, Vol. 1.* New York: Wiley & Sons.

Carlson, M., Marcus-Newhall, A., & Miller, N. (1990). Effects of situational aggression cues: A quantitative review. *Journal of Personality and Social Psychology, 58*, 622–633.

Caspi, A., Henry, B.C., McGee, R.O., Moffitt, T.E., & Silva, P.A. (1995). Temperamental origins of child and adolescent behavior problems: From age three to age fifteen. *Child Development, 66*, 55–68.

Caspi, A., Moffitt, T.E., Morgan, J., Rutter, M., Taylor, A., Arseneault, L.. Tully, L., Jacobs, C., Kim-Cohen, J., Polo-Tomas, M. (2004). Maternal Expressed Emotion Predicts Children's Antisocial Behavior Problems: Using Monozygotic-Twin Differences to Identify Environmental Effects on behavioral development. *Developmental Psychology. 40*, 149–161.

Cassidy, J., Kirsh, S.J., Scolton, K.L., & Parke, R.D. (1996). Attachment and representations of peer relationships. *Developmental Psychology, 32,* 892–904.

Chang, L., Schwartz, D., Dodge, K.A., & McBride-Chang, C. (2003). Harsh parenting in relation to child emotion regulation and aggression. *Journal of Family Psychology, 17,* 598–606.

Cleckley, H. (1976). *The Mask of Sanity.* St. Louis, MO: Mosby.

Cicchetti, D., & Cohen, D.J. (eds.). (1995). Preface. *Developmental Psychopathology, Vol. 1: Theories and Methods.* Canada: Wiley & Sons.

Cicchetti, D., & Toth, S.L. (2006). Developmental Psychopathology and Preventive Intervention. *Handbook of Child Psychology, 6th ed: Vol. 4, Child Psychology in Practice* (pp. 497–547). Hoboken, NJ: John Wiley & Sons.

Cicchetti, D., & Crick, N.R. (2009). Editorial. Precursors and diverse pathways to personality disorder in children and adolescents. *Development and Psychopathology, 21,* 683–685.

Cicchetti D., & Curtis, W.J. (eds.). (2007). A multilevel approach to resilience. *Development and Psychopathology, 19,* 627–955.

Clawson, E. (2009). The new canaries in the mine: The priority of human welfare in animal abuse prosecution. In A. Lindzey (ed.), *The Link between Animal Abuse and Human Violence* (pp. 190–200). Sussex: Academic Press.

Cloninger. C.R., Dragan, M.S., & Thomas R.P. (1993). A psychobiological model of temperament and character. *Archives of General Psychiatry 50,* 975–990.

Conduct Problems Prevention Research Group. (1992). A developmental and clinical model for the prevention of conduct disorder: The FAST Track Program. *Development and Psychopathology, 4,* 509–527.

Coston, C., & Protz, C.M. (1998). Kill your dog, beat your wife, screw your neighbour's kids, rob a bank? A cursory look at an individuals' vat of social chaos resulting from deviance. *Free Inquiry in Creative Sociology, 26,* 153–158.

Crick, N.R., Casas, J.F., & Mosher, M. (1997). Relational and overt aggression in preschool. *Developmental Psychology. 33,* 579–588.

Crick, N.R., & Dodge, K.A. (1994). A review and reformulation of social information processing mechanisms in children's adjustment. *Psychological Bulletin, 115,* 74–101.

Crick, N.R., & Grotpeter, J.K. (1995). Relational aggression, gender and social-psychological adjustment. *Child Development, 66,* 10–22.

Crick, N.R., & Werner, N.E. (1998). Response decision processes in relational and overt aggression. *Child Development. 69,*1630–1639.

Crijnen, A.M., Achenbach, T.M., Verhulst, F.C. (1997). Comparisons of problems reported by parents of children in 12 cultures: Total problems, externalizing, and internalizing. *Journal of the American Academy of Child & Adolescent Psychiatry, 36,* 1269–1277.

Cummings, E.M. (1987). Coping with background anger in early childhood. *Child Development, 58,* 976–984.

Cunningham, A., & Baker, L. (2004). *What about Me! Seeking to Understand a Child's View of Violence in the Family.* London: Centre for Children & Families in the Justice System.

Cummings, E.M. (1987). Coping with background anger in early childhood. *Child Development, 58,* 976–984.

Currie, C.L. (2006). Animal cruelty by children exposed to domestic violence. *Child Abuse & Neglect, 30,* 425–435.

Dadds, M.R. (2008). Conduct problems and cruelty to animals in children: What is the link? In F.R. Ascione (ed.). *The International Handbook of Animal Abuse and Cruelty: Theory, Research, and Application* (pp. 111–131). West Lafayette, Indiana: Purdue University Press.

Dadds, M.R., Turner, C.M., & McAloon, J. (2002). Developmental links between cruelty to animals and human violence. *Australian & New Zealand Journal of Counselling, 35,* 363–382.

Dadds, M.R., Whiting, C., & Hawes, D.J. (2006). Associations among cruelty to animals, family conflict, and psychopathic traits in childhood. *Journal of Interpersonal Violence, 21,* 411–429.

Daniell, C. (2001). Ontario's SPCA's women's shelter survey shows staggering results. *The Latham Letter,* Spring, 16–17.

Davies, P.T., & Cummings, E. (2006). Interparental discord, family process, and developmental psychopathology. In D. Cicchetti & D. J. Cohen (eds.), *Developmental psychopathology, Vol. 3: Risk, disorder, and adaptation* (2nd ed., pp. 86–128). Hoboken, NJ: John Wiley & Sons.

Davies, P.T., Myers, R.L., Cummings, E.M., & Heindel, S. (1999). Adult conflict history and children's subsequent responses to conflict: An experimental test. *Journal of Family Psychology, 13,* 610–628.

Deater-Deckard, K., & Dodge, K.A. (1997). Spare the rod, spoil the authors: Emerging themes in research on parenting and child development. *Psychological Inquiry. 8,* 230–235.

Deater-Deckard, K., Dodge, K.A., Bates, J.E., Pettit, G.S. (1996). Physical discipline among African American and European American mothers: Links to children's externalizing behaviors. *Developmental Psychology. 32,* 1065–1072.

DeGue, S., & DiLillo, D. (2009). Is animal cruelty a "red flag" for family violence?: Investigating co-occurring violence toward children, partners, and pets. *Journal of Interpersonal Violence, 24,* 1036–1056.

Deluty, D.J. (1983). Children's evaluations of aggressive, assertive, and submissive responses. *Journal of Clinical Child Psychology, 12,* 124–129.

Deviney, E., Dickert, J., & Lockwood, R. (1983). The care of pets within child abusing families. *International Journal for the Study of Animal Problems, 4,* 321–329.

DeWall, C.N., & Anderson, C.A. (2011). The General Aggression Model. In P.R. Shaver & M. Mikulincer, M. (eds.), *Human Aggression and Violence: Causes, manifestations, and consequences.* (pp. 15–33). Washington, D.C.: American Psychological Association.

Dishion, T.J., French, D.C., & Patterson, G.R. (1995). The development and ecology of antisocial behavior. In D. Cicchetti & D.J. Cohen (eds.), *Developmental Psychopathology, Vol. 2: Risk, Disorder, and Adaptation* (pp. 421–471). Oxford, England: John Wiley & Sons.

Dodge, K.A. (1983). Behavioral antecedents of peer status. *Child Development, 54,* 1386–1399.

Dodge, K.A. (2011). Social information processing patterns as mediators of the interaction between genetic factors and life experiences in the development of aggressive behavior. In P.R. Shaver & M. Mikulincer, M. (eds.), *Human Aggression and Violence: Causes, Manifestations, and Consequences.* (pp. 165–185). Washington, D.C.: American Psychological Association.

Dodge, K.A., Bates, J.E., & Pettit, G.S. (1990). Mechanisms in the cycle of violence. *Science, 250,* 1678–1683.

Dodge, K.A., & Coie, J.D. (1987). Social-information-processing factors in reactive and proactive aggression in children's peer groups. *Journal of Personality and Social Psychology, 53,* 1146–1158.

Dodge, K.A., Coie, J.D., & Lynam, D. (2006). Aggression and antisocial behavior in youth. In N. Eisenberg, W. Damon & R. M. Lerner (eds.), *Handbook of child psychology* (6th ed., Vol. 3, Social, emotional, and personality development, pp. 719–788). Hoboken, NJ: John Wiley & Sons.

Dodge, K.A., Murphy, R.R., & Buchsbaum, K. (1984). The assessment of intention-cue detection skills in children: Implications for developmental psychopathology. *Child Development, 55,* 163–173.

Dodge, K.A., Pettit, G.S., & Bates, J.E. (1994). Socialization mediators of the relation between socioeconomic status and child conduct problems. *Child Development. 65,* 649–665.

Dodge, K.A., Pettit, G.S., McClaskey, C.L., Brown, M.M., & Gottman, J.M. (1986). Social Competence in Children. *Monographs of the Society for Research in Child Development, 51,* No. 2.

Dodge, K.A., Price, J.M., Bachorowski, J.A., & Newman, J.P. (1990). Hostile attributional biases in severely aggressive adolescents. *Journal of Abnormal Psychology, 99,* 385–392.

Dodge, K.A., & Tomlin, A. (1987). Utilization of self-schemas as a mechanism of attributional bias in aggressive children. *Social Cognition, 5,* 280–300.

Dollard, J., Doob, L.W., Miller, N.E., Mower, O.H., & Sears, R.R. (1939). *Frustration and Aggression.* New Haven, CT: Yale University Press.

Duncan, A., & Miller, C. (2002). The impact of an abusive family context on childhood animal cruelty and adult violence. *Aggression and Violent Behavior, 7,* 365–383.

Duncan, A., Thomas, J.C., & Miller, C. (2005). Significance of family risk factors in development of childhood animal cruelty in adolescent boys with conduct problems. *Journal of Family Violence, 20,* 235–239.

Dutton, D.G. (2000). Witnessing parental violence as a traumatic experience shaping the abusive personality. *Journal of Aggression, Maltreatment and Trauma, 3,* 59–67.

Egan, S., Monson, T., & Perry, D. (1998). Social-cognitive influences on change in aggression over time. *Developmental Psychology, 34,* 996–1006.

Egeland, B. & Sroufe, L.A. (1981) Attachment and early maltreatment *Child Development, 52,* 44–52.

Eisenberg, N., Champion, C., & Ma, Y. (2004). Emotion-related regulation: An emerging construct. *Merrill-Palmer Quarterly, 50,* 236–259.

Eisenberg, N., & Fabes, R.A. (1999). Emotion, emotion-related regulation, and quality of socioemotional functioning. In L. Balter & C.S. Tamis-LeMonda (eds.), *Child Psychology: a Handbook of Contemporary Issues.* (pp. 318–335), New York: Psychology Press.

Eisenberg, N., Valiente, C., Morris, A.S., Fabes, R.A., Cumberland, A., Reiser, M., et al. (2003). Longitudinal relations among parental emotional expressivity, children's regulation, and quality of socioemotional functioning. Developmental Psychology, 39, 3–19.

Eisenberg, N., Zhou, Q., Spinrad, T.L., Valiente, C., Fabes, R.A., & Liew, J. (2005). Relations among positive parenting, children's effortful control, and externalizing problems: A three-wave longitudinal study. *Child Development, 76,* 1055–1071.

Elliot, D.S. (1994). Serious violent offenders: Onset, developmental course, and termination – The American Society of Criminology 1193 Presidential Address. *Criminology, 32,* 1–21.

Erdley, C.A., Asher, S.R. (1998). Linkages between children's beliefs about the legitimacy of aggression and their behaviour. *Social Development, 7,* 321–339.

Eron, L.D. (1987). The development of aggressive behaviour from the perspective of a developing behaviorist. *American Psychologist, 42,* 435–442.

Eron, L. (2001). Seeing is believing: How viewing violence alters attitudes and aggressive behavior. In A.C. Bohart & D.J. Stipek (eds.), *Constructive & Destructive Behavior: Implications for Family, School, & Society* (pp. 49–60). Washington, D.C.: American Psychological Association.

Eron, L.D., Huesmann, L.R., Lefkowitz, M.M., & Walder, L.O. (1972). Does television violence cause aggression? *American Psychologist. 27,* 253–263.

Eysenck, H.J. (1967). *The Biological Basis of Personality.* Baltimore, Maryland: University Park Press.

Eysenck, H.J., & Eysenck, S.B.G. (1975). *Manual of the Eysenck Personality Questionnaire (Junior and adult).* Kent: Hodder and Stoughton.

Fabes, R.A., & Eisenberg, N. (1992). Young children's emotional arousal and anger/aggressive behaviours. In A. Fraezek & H. Zumkley (eds.), *Socialization and Aggression* (pp. 85–102). Berlin, Germany: Springer-Verlag.

Farrington, D.P. (1978). The family background of aggressive youths. In L.A. Hersov, M. Berger, & D. Schaffer (eds.), *Aggression and Antisocial Behavior in Childhood and Adolescence.* Oxford: Pergamon.

Farrington, D.P. (1991). Childhood aggression and adult violence: Early precursors and later life outcomes. In D.J. Peplar & H.K. Rubin (eds.), *The Development and Treatment of Childhood Aggression* (pp. 5–29). Hillsdale, N.J.: Erlbaum.

Farrington, D.P. (1994). Childhood, adolescent, and adult features of violent males. In L.R. Huesmann (ed.), *Aggressive Behavior: Current Perspectives.* (pp. 215–240). New York: Plenum.

Farrington, D.P. (1995). The development of offending and antisocial behaviour from childhood: Key findings from the Cambridge study in delinquent development. *Journal of Child Psychology and Psychiatry, 36,* 1–36.

Farrington, D.P. & Hawkins, J.D. (1991). Predicting participation, early onset and later persistence in officially recorded offending. *Criminal Behaviour and Mental Health. 1,* 1–33.

Faver, C.A. (2010). School-based humane education as a strategy to prevent violence: Review and recommendations. *Children and Youth Services Review, 32*(3), 365–370.

Faver, C.A., & Cavazos, A.M. (2007). Animal abuse and domestic violence: A view from the border. *Journal of Emotional Abuse, 7,* 59–81.

Felthous, A.R. (1980). Aggression against cats, dogs and people. *Child Psychiatry & Human Development, 10*(3), 169–177.

Felthous, A.R., & Bernard, H. (1979). Enuresis, firesetting, and cruelty to animals. *Journal of Forensic Science, 24,* 240–246.

Felthous, A.R., & Kellert, S.R. (1986). Violence against animals and people: Is aggression against living creatures generalized? *Bulletin of the American Academy of Psychiatry & the Law, 14*, 55–69.

Felthous, A.R., & Kellert, S.R. (1987). Childhood cruelty to animals and later aggression against people: A review. *American Journal of Psychiatry, 144*(6), 710–717.

Felthous, A.R., & Yudowitz, B. (1977). Approaching a comparative typology of assaultive female offenders, *Psychiatry, 40*, 270–276.

Fergusson, D.M., & Horwood, L.J. (1998). Exposure to interparental violence in childhood and psychosocial adjustment in young adulthood. *Child Abuse and Neglect, 22*, 339–357.

Feshbach, S. (1964). The function of aggression and the regulation of aggressive drive. *Psychological Review, 71*, 257–272.

Flynn, C.P. (1999a). Animal abuse in childhood and later support for interpersonal violence in families. *Society and Animals, 7*, 161–172.

Flynn, C.P. (1999b). Exploring the link between corporal punishment and children's cruelty to animals. *Journal of Marriage & the Family, 61*, 971–981.

Flynn, C.P. (2000). Why family professionals can no longer ignore violence toward animals. *Family Relations: Interdisciplinary Journal of Applied Family Studies, 49*, 87–95.

Flynn, C.P. (2011). Examining the links between animal abuse and human violence. *Crime, Law and Social Change, 55*, 453–468.

Fraser, M.W., Galinsky, M. J., Smokowski, P.R., Day, S. H., Terzian, M.A., Rose, R.A., et al. (2005). Social Information-Processing Skills Training to Promote Social Competence and Prevent Aggressive Behavior in the Third Grades. *Journal of Consulting and Clinical Psychology, 73*, 1045–1055.

Frick, P. J. (2002). Juvenile psychopathy from a developmental perspective: Implications for construct development and use in forensic assessments. *Law and Human Behavior, 26*(2), 247–253.

Frick, P.J., & Dickens, C. (2006). Current perspectives on conduct disorder. *Current Psychiatry Reports, 8*, 59–72.

Frick, P.J., Lahey, B.B., Loeber, R., Tannenbaum, L., & et al. (1993). Oppositional defiant disorder and conduct disorder: A meta-analytic review of factor analyses and cross-validation in a clinic sample. *Clinical Psychology Review, 13*, 319–340.

Frick, P.J., O' Brien, B.S., Wootton, J.M., & McBurnett, K. (1994). Psychopathy and conduct problems in children. *Journal of Abnormal Psychology, 103*, 700–707.

Frick, P.J., & Viding, E. (2009). Antisocial behaviour from a developmental psychopathology perspective. *Development and Psychopathology, 21*, 1111–1131.

Frick, P.J., & White, S.F. (2008). Research review: The importance of Callous-Unemotional traits for developmental models of aggressive and antisocial behavior. *Journal of Child Psychology and Psychiatry, 49*, 359–375.

Gaensbauer, T.J., Mrazek, D., & Harmon, R.J. (1981). Emotional expression in abused and/or neglected infants. In N. Frude (ed.), *Psychological approaches to child abuse* (pp. 120–135). Totowa, N.J.: Rowan and Littlefield.

Gelhorn, H.L., Sakai, J.T., Price, R.K., & Crowley, T.J. (2007). DSM-IV conduct disorder criteria as predictors of antisocial personality disorder. *Comprehensive Psychiatry, 48*, 529–538.

150   *References*

Gendreau, P.L., & Archer, J. (2005). Subtypes of aggression in humans and animals. In R.E. Tremblay, W.W. Hartup, & J. Archer (eds.), *Developmental Origins of Aggression* (pp. 25–45). New York: Guilford Press.

Gershoff, E.T. (2002). Parental corporal punishment and associated child behaviours and experiences: A meta-analytic and theoretical review. *Psychological Bulletin, 128,* 539–579.

Gifford-Smith, M.E., & Rabiner, D.L. (2004). Social information processing and children's social adjustment. In J. Kupersmidt & K.A. Dodge (eds.), *Children's Peer Relations: from Development to Intervention* (pp. 69–84). Washington, D.C.: American Psychological Association.

Gilliom, M., Shaw, D.S., Beck, J.E., Schonberg, M.A., & Lukon, J.L. (2002). Anger regulation in disadvantaged preschool boys: Strategies, antecedents, and the development of self-control. *Developmental Psychology, 38,* 222–235.

Gleyzer, R., Felthous, A.R., & Holzer, C.E. (2002). Animal cruelty and psychiatric disorders. *Journal of the American Academy of Psychiatry & the Law, 30,* 257–265.

Glueck, S. (1959). *The Problem of Delinquency.* Oxford, England: Houghton-Mifflin.

Glueck, S., & Gluek, E. (1950). *Unraveling Juvenile Delinquency.* Oxford, England: Commonwealth Fund.

Gordon R.G. (1939) *Survey of Child Psychiatry.* Oxford, England: Oxford University Press.

Gouze, K.R. (1987). Attention and social problem solving as correlates of aggression in preschool males. *Journal of Abnormal Child Psychology, 15,* 181–197.

Graham-Bermann, S.A., & Hughes, H.M. (1998). The impact of domestic violence and emotional abuse on children: The intersection of research, theory, and clinical intervention. *Journal of Emotional Abuse, 1,* 1–21.

Graham-Bermann, S.A., & Levendosky, A.A. (1998). Traumatic stress symptoms in children of battered women. *Journal of Interpersonal Violence, 13,* 111–128.

Grant, B.F., Harford, D.A., Dawson, D.A., & Pickering, R.P. (1995). The Alcohol Use Disorder and Associated Disabilities Interview Schedule (AUDADIS): Reliability of alcohol and drug modules in a general population sample. *Drug and Alcohol Dependence, 39,* 37–44.

Green, A.E., Gestern, E.L., Greenwald, M.A., & Salcedo, O. (2008). Predicting delinquency in adolescence and young adulthood: A longitudinal analysis of early risk factors. *Youth Violence and Juvenile Justice, 6,* 323–342.

Greenberg, M.T., Speltz, M.L., & DeKlyen, M. (1993). The role of attachment in the early development of disruptive behaviour problems. *Development and Psychopathology, 5,* 191–213.

Greeson, L.E., & Williams, R.A. (1986). Social implications of music videos for youth: An analysis of the content and effects of MTV. *Youth and Society, 18,* 177–189.

Griffin, R.S., & Gross, A.M. (2004). Childhood bullying: Current empirical findings and future directions for research. *Aggression and Violent Behavior, 9,* 379–400.

Guerra, N.G., Huesmann, L.R., & Spindler, A. (2003). Community violence exposure, social cognition, and aggression among urban elementary school children. *Child Development, 74,* 1561–1576.

Guerra, N.G., Huesmann, L.R., Tolan, P.H., Van Acker, R., Eron, L.D. (1995). Stressful events and individual beliefs as correlates of economic disadvantage and aggression among urban children. *Journal of Consulting and Clinical Psychology, 63,* 518–528.

Gullone, E. (2000). The development of normal fear: A century of research. *Clinical Psychology Review. 20 (4),* 429–451.

Gullone, E. (2009). A lifespan perspective on human aggression and animal abuse. In A. Lindzey (ed.), *The Link between Animal Abuse and Human Violence* (pp. 38–60). Sussex: Academic Press.

Gullone, E., & Clarke, J. (2008). Human-Animal Interactions: The Australian Perspective. In F. Ascione (ed.), *The International Handbook of Theory and Research on Animal Abuse and Cruelty* (pp. 305–335). West Lafayette, Indiana: Purdue University Press.

Gullone, E., Hughes, E.K., King, N.J., & Tonge, B. (2010). The normative development of emotion regulation strategy use in children and adolescents: A 2-year follow-up study. *Journal of Child Psychology and Psychiatry. 51,* 567–674.

Gullone, E., & Robertson, N. (2008). The relationship between bullying and animal abuse in adolescents: The importance of witnessing animal abuse. *Journal of Applied Developmental Psychology, 29,* 371–379.

Gumpel, T.P., & Meadan, H. (2000). Children's perceptions of school-based violence. *British Journal of Educational Psychology, 70* (Pt 3), 391–404.

Guymer, E., Mellor, D., Luk, E., & Pearse, V. (2001). The development of a screening questionnaire for childhood cruelty to animals. *Journal of Child Psychology and Psychiatry, 42,* 1057–1063.

Haapasalo, J., Tremblay, R.E. (1994). Physically aggressive boys from ages 6 to 12: Family background, parenting behavior, and prediction of delinquency. *Journal of Consulting and Clinical Psychology, 62,* 1004–1052.

Halberstadt, A.G., Denham, S.A., & Dunsmore, J.C. (2001). Affective social competence. *Social Development, 10,* 79–119.

Hansen, C.H., & Hansen, R.D. (1990). Rock music videos and antisocial behavior. *Basic and Applied Social Psychology, 11,* 357–369.

Hare, R.D. (1978). Electrodermol and cardiovascular correlates of psychopathy. In R.D. Hare and D. Schalling (eds.), *Psychopathic Behavior: Approaches to Research* (pp. 107–144). New York: Wiley.

Hare, R.D. (1993). Without conscience: The disturbing world of the psychopaths among us. New York: Guilford Press.

Hare, R.D. (2003). *Manual for the Revised Psychopathy Checklist* (2nd ed.). Toronto: Multi-Health Systems.

Hare, R.D., Hart, S.D., & Harpur, T.J. (1991). Psychopathy and the DSM–IV Criteria for Antisocial Personality Disorder. *Journal of Abnormal Psychology. 100,* 391–398.

Hare, R.D., McPherson, L.M., & Forth, A.E. (1988). Male psychopaths and their criminal careers. *Journal of Consulting and Clinical Psychology, 56,* 710–714.

Harpur, T.J., Hare, R.D., & Hakstian, A.R. (1989). Two-Factor Conceptualization of Psychopathy: Construct Validity and Assessment Implications. *Psychological Assessment: A Journal of Consulting and Clinical Psychology. 1,* 6–17.

Hartup, W.W. (2005). The development of aggression. In R.E. Tremblay, W.W. Hartup, & J. Archer (eds.), *Developmental Origins of Aggression* (pp. 3–22). New York: Guilford Press.

Haskett, M.E., & Kistner, J.A. (1991). Social interactions and peer perceptions of young physically abused children. *Child Development, 62,* 979–990.

Hastings, P.D., Zahn-Waxler, C., Robinson, J., Usher, B., & Bridges, D. (2000). The development of concern for others in children with behavior problems. *Developmental Psychology, 36,* 531–546.

Hay, D.F., Nash, A., & Pederson, J. (1983). Interactions between 6-month olds. *Child Development, 54,* 557–562.

Heath, G.A., Hardesty, V.A., & Goldfine, P.E. (1984). Firesetting, enuresis, and animal cruelty. *Journal of Child and Adolescent Psychotherapy, 1,* 97–100.

Heller, M.S., Ehrlich, S.M., & Lester, D. (1984). Childhood cruelty to animals, firesetting, and enuresis as correlates of competence to stand trial. *Journal of General Psychology, 110,* 151–153.

Hellman, D.S., & Blackman, N. (1966). Enuresis, firesetting, and cruelty to animals: A triad predictive of adult crime. *American Journal of Psychiatry, 122,* 1431–1435.

Henry, B.C. (2004a). Exposure to animal abuse and group context: Two factors affecting participation in animal abuse. *Anthrozoos, 17,* 290–305.

Henry, B.C. (2004b). The relationship between animal cruelty, delinquency, and attitudes toward the treatment of animals. *Society & Animals, 12,* 185–207.

Henry, B., Caspi., Moffitt, T.E., & Silva, P.W. (1996). Temperamental and familial predictors of violent and nonviolent criminal convictions: From age 3 to 18. *Developmental Psychology, 32,* 614–623.

Henry, B.C., & Sanders, C.E. (2007). Bullying and animal abuse: Is there a connection? *Society & Animals, 15,* 107–126.

Hensley, C., & Tallichet, S. E. (2005). Animal cruelty motivations: Assessing demographic and situational influences. *Journal of Interpersonal Violence, 20,* 1429–1443.

Hensley, C., Tallichet, S.E., & Dutkiewicz, E.L. (2009). Recurrent childhood animal cruelty: Is there a relationship to adult recurrent interpersonal violence? *Criminal Justice Review, 34,* 248–257.

Herrenkhohl, T.I., Maguin, E., Hill, K.G., Abbott, R.D., & Catalano, R.F. (2000). Developmental risk factors for youth violence. *Journal of Adolescent Health, 26,* 176–186.

Huesmann, L. (1986). Psychological processes promoting the relation between exposure to media violence and aggressive behavior by the viewer. *Journal of Social Issues, 42,* 125–139.

Huesmann, L. (1988). An information processing model for the development of aggression. *Aggressive Behavior, 14,* 13–24.

Huesmann, L. (1998). The role of social information processing and cognitive schemas in the acquisition and maintenance of habitual aggressive behaviour. In R.G. Geen and E. Donnerstein (eds.), *Human Aggression: Theories, Research and Implications for Policy.* (pp. 73–109). New York: Academic Press.

Huesmann, L.R., Eton, L.D., Lefkowitz, M.M. & Walder, L.O. (1984). Stability of aggression across time and generations. *Developmental Psychology, 20,* 1120–1134.

Huesmann, L.R., & Guerra, N.G. (1997). Children's normative beliefs about aggression and aggressive behaviour. *Journal of Personality and Social Psychology. 72,* 408–419.

Huesmann, L.R., & Moise, J. (1998). The stability and continuity of aggression from childhood to young adulthood. In D.J. Flannery & C.R. Huff (eds.), *Youth*

Violence: Prevention, Intervention, and Social Policy. (pp. 73–95). Washington, D.C.: American Psychiatric Press.

Howard, R.C. (1984). The clinical EEG and personality in mentally abnormal offenders. *Psychological Medicine, 14,* 569–580.

Howell, J.C., Krisberg, B., & Jones, M. (1995). Trends in juvenile crime and youth violence. In J.C. Howell, B. Krisberg, J.D. Hawkins, & J.J. Wilson (eds.), *Serious, Violent, and Chronic Juvenile Offenders* (pp. 1–35). Thousand Oaks, CA: Sage.

Humane Education Society, United Kingdom (2010). http://www.humaneeducationsociety.co.uk/humane-education-society Retrieved 1st July, 2010.

Humane Society of the United States (1997). http://www.hsus.org/firststrike/factsheets/index.html Retrieved 17th July, 2010.

Huesmann, L.R., Eron, L.D., Lefkowitz, M.M., & Walder, L.O. (1984). Stability of aggression over time and generations. *Developmental Psychology, 20,* 1120–1134.

Huesmann, L.R., Miller, L.S. (1994). Long-term effects of repeated exposure to media violence in childhood. In L. Huesmann (ed.). *Aggressive Behavior: Current Perspectives.* (pp. 153–186). New York: Plenum Press.

Huesmann, L.R., & Moise, J. (1998). The stability and continuity of aggression from early childhood to young adulthood. In D.J. Flannery & C.R. Huff (eds.), *Youth Violence: Prevention, Intervention, and Social Policy* (pp. 73–95). Washington, D.C.: American Psychiatric Press.

Huesmann, L.R., Moise-Titus, J., Podolski, C.L., & Eron, L.D. (2003). Longitudinal relations between children's exposure to TV violence and their aggressive and violent behaviour in young adulthood: 1977–1992. *Developmental Psychology. 39,* 201–221.

Hutton, J.S. (1983). Animal abuse as a diagnostic approach in social work: A pilot study (reprinted). In R. Lockwood & F.R. Ascione (eds.) (1998). *Cruelty to Animals and Interpersonal Violence: Readings in Research and Application* (pp. 415–420). West Lafayette, IN: Purdue University Press.

Jaffee, S.R., Caspi, A., Moffitt, T.E., & Taylor, A. (2004). Physical maltreatment victim to antisocial child. Evidence of an environmentally mediated process. *Journal of Abnormal Psychology, 113,* 44–55.

Johnson, J.G., Cohen, P., Smailes, E.M., Kasen, L., & Brook, (2002). Television viewing and aggressive behavior during adolescence and adulthood. *Science, 295,* 2468–2471.

Jouriles, E.N., Barling, J., & O ' Leary, K.D. (1987). Predicting child behavior problems in maritally violent families. *Journal of Abnormal Child Psychology, 15,* 165–173.

Kaplan, H.B., & Liu, X. (1994). A longitudinal analysis of mediating variables in the drug use-dropping out relationship. *Criminology, 32,* 415–439.

Kagan, J., & Snidman, N. (1991). Temperamental factors in human development. *American Psychologist, 46*(8), 856–862.

Karasaki, T.S.J. (1999). A pet's life in Japan – unloved, abandoned, destroyed. Revised law to designed to give pets better lives. Ashai Evening News. Available at http://www.aapn.org/japandpgs.html. Retrieved 12 November 2010.

Katz, L.F., & Gottman, J. (1993). Patterns of marital conflict predict children's internalizing and externalizing disorders. *Developmental Psychology, 29,* 940–950.

Kaufman, J., & Cicchetti, D. (1989). Effects of maltreatment on school-age children's socioemotional development: Assessments in a day-camp setting. *Developmental Psychology, 25,* 516–524.

Kaufman, J., & Zigler, E. (1987). Do abused children become abusive parents? *American Journal of Orthopsychiatry, 57,* 186–192.

Kazdin, A.E. (1987). Treatment of antisocial behavior in children: Current status and future directions. *Psychological Bulletin, 102,* 187–203.

Keenan, K., Loeber, R., Zhang, Q., & Stouthamer-Loeber, M. (1995). The influence of deviant peers on the development of boys' disruptive and delinquent behavior: A temporal analysis. *Development and Psychopathology. 7,* 715–726.

Keenan, K., & Shaw, D. (1997). Developmental and social influences on young girls' early problem behavior. *Psychological Bulletin, 121,* 95–113.

Keenan, K., & Shaw, D. (2003). Starting at the beginning: Exploring the etiology of antisocial behavior in the first year of life. In B.B. Lahey, T.E. Moffitt, & A. Caspi (eds.), *Causes of conduct disorder and juvenile delinquency* (pp. 153–181). New York: Guilford Press.

Keenan, K., Shaw, D.S., Delliquadri, E., Giovanelli, J., & Walsh, B. (1998). Evidence for the continuity of early problem behaviours: Application of a developmental model. *Journal of Abnormal Child Psychology, 26,* 441–452.

Keeney, B.T., & Heide, K.M. (1995). Serial murder: A more accurate and inclusive definition. *International Journal of Offender and Comparative Criminology, 39,* 299–306.

Kellam, S.G., Ling, X., Merisca, R., Brown, C.H., & Ialongo, N. (1998). The effect of the level of aggression in the first grade classroom on the course and malleability of aggressive behavior into middle school. *Development and Psychopathology. 10,* 165–185.

Kellert, S.R., & Felthous, A.R. (1985). Childhood cruelty toward animals among criminals and noncriminals. *Human Relations, 38*(12), 1113–1129.

Keltner, D., & Robinson, R.J. (1996). Extremism, power, and the imagined basis of social conflict. *Current Directions in Psychological Science, 5,* 101–105.

Kingston, L., & Prior, M. (1995). The development of patterns of stable, transient, and school-age onset aggressive behavior in young children. *Journal of the American Academy of Child and Adolescent Psychiatry, 34,* 348–358.

Kochanska, G. (1993). Toward a synthesis of parental socialisation and child temperament in early development of conscience. *Child Development, 64,* 325–347.

Krueger, R.F. (1999). Personality traits in late adolescence predict mental disorders in early adulthood: A prospective-epidemiological study. *Journal of Personality, 67,* 39–65.

Krueger, R.F., Caspi, A., Moffit, T.E., White, J., & Stouthamer-Loeber, M. (1996). Delay of gratification, psychopathology, and personality: Is low self-control specific to externalizing problems? *Journal of Personality, 64,* 107–129.

Krug, E.G., Dahlberg, L.L., Mercy, J.A., Zwi, A.B., & Lozano, R. (2002). World report on violence and health. Geneva: World Health Organization.

Kupersmidt, J.B., & Coie, J.D. (1990). Preadolescent peer status, aggression, and school adjustment as predictors of externalizing problems in adolescence. *Child Development, 62,* 1350–1362.

Lahey, B.B., & Loeber, R. (1994). Framework for a developmental model of oppositional defiant disorder and conduct disorder. In D. Routh (ed.), *Disruptive Behavior Disorders in Childhood* (pp. 139–180). New York: Plenum Press.

Lahey, B.B., Loeber, R., Burke, J.D., & Applegate, B. (2005). Predicting future antisocial personality disorder in males from a clinical assessment in childhood. *Journal of Consulting and Clinical Psychology, 73,* 389–399.

Laible, D.J., & Thompson, R.A. (1998). Attachment and emotional understanding in preschool children. *Developmental Psychology, 34,* 1038–1045.

Lansford, J.A., Deater-Deckard, K., Dodge, , K.A., Bates, J.E., & Pettit, G.S. (2004). Ethnic differences in the link between physical discipline and later adolescent externalizing behaviours. *Journal of Child Psychology and Psychiatry,* 801–812.

Lansford, J.A., Dodge, K.A., Pettit, G.S., Bates, J.E., Crozier, J., & Kaplow, J. (2002). A 12-year prospective study of the long-term effects of early child physical maltreatment on psychological, behavioral, and academic problems in adolescence. *Archives of Pediatrics and Adolescent Medicine, 156,* 824–830.

Laperspetz, K.M., & Bjorkqvist, K. (1992). Indirect aggression in girls and boys. In L.R. Huesmann (ed.), *Aggressive Behavior: Current Perspectives.* (pp. 131–150). New York: Plenum.

Larsson, H., Viding, E., Rijsdijk, F.V., & Plomin, R. (2008). Relationships between parental negativity and childhood antisocial behavior over time: A bidirectional effects model in a longitudinal genetically informative design. *Journal of Abnormal Child Psychology, 36,* 633–645.

Lemerise, E.A., & Arsenio, W.F. (2000). An Integrated Model of Emotion Processes and Cognition in Social Information Processing. *Child Development.* 107–118.

Lenzenweger, M.F., Lane, M.C., Loranger, A.W., & Kessler, R.C. (2007). DSM-IV personality disorders in the National Comorbidity Survey Replication. *Biological Psychiatry, 62,* 553–564.

Levinson, B.M. (1972). *Pets and human development.* Springfield: C.C. Thomas.

Linzey, A. (2009a). Introduction. Does abuse really benefit us? In A. Lindzey (ed.), *The Link between Animal Abuse and Human Violence* (pp. 1–10). Sussex, Academic Press.

Linzey, A. (2009b). *Why Animal Suffering Matters.* Oxford: Oxford University Press.

Locke, J. (1989). *Some Thoughts Concerning Education* (5th ed.) (J.W. & J.S., Eds), Oxford: Clarendon Press. (Original work published 1693).

Lockwood, R. (2008). Counting cruelty: Challenges and opportunities in assessing animal abuse and neglect in America. In F.R. Ascione (ed.). *The International Handbook of Animal Abuse and Cruelty: Theory, Research, and Application* (pp. 87–109). West Lafayette, Indiana: Purdue University.

Lockwood, R., & Church, A. (1996). Deadly serious: An FBI perspective on animal cruelty. *HSUS News* (Fall), 27–30.

Lockwood, R., & Hodge, G.R. (1986; 1998). The tangled web of animal abuse: The links between cruelty to animals and human violence. In Lockwood, R., & Ascione, F.R. *Cruelty to Animals and Interpersonal Violence: Readings in Research and Application.* (pp. 77–82). West Lafayette, Indiana: Purdue University Press.

Loeber, R. (1982). The stability of antisocial and delinquent child behavior: A review. *Child Development, 53,* 1431–1446.

Loeber, R. (1985). Patterns and development of antisocial child behaviour. *Annals of Child Development, 2,* 77–116.

Loeber, R., & Dishion, T. (1983). Early predictors of male delinquency: A review. *Psychological Bulletin, 93,* 68–99.

Loeber, R., Farrington, D.P., Stouthamer-Loeber, M., & van Kammen, W.B. (1998). *Antisocial Behaviour and Mental Health Problems: Explanatory Factors in Childhood and Adolescence.* Mahwah, Pittsburgh, PA: Erlbaum.

Loeber, R., & Hay, D. (1997). Key issues in the development of aggression and violence from childhood to early adulthood. *Annual Review of Psychology, 48,* 371–410.

Loeber, R., Lahey, B.B., & Thomas, C. (1991). Diagnostic conundrum of oppositional defiant disorder and conduct disorder. *Journal of Abnormal Psychology, 100,* 379–390.

Loring, M. T., & Bolden-Hines, T.A. (2004). Pet Abuse by Batterers as a Means of Coercing Battered Women Into Committing Illegal Behavior. *Journal of Emotional Abuse, 4,* 27–37.

Luk, E.S., Staiger, P. K., Wong, L., & Mathai, J. (1999). Children who are cruel to animals: A revisit. *Australian and New Zealand Journal of Psychiatry, 33,* 29–36.

Luntz, B.K. & Widom, C.S. (1994). Antisocial Personality Disorder in Abused and Neglected Children Grown Up. *The American Journal of Psychiatry, 151,* 670–674.

Lynam, D.R. (1996). Early identification of chronic offenders: Who is the fledgling psychopath? *Psychological Bulletin, 120,* 209–234.

Lynam, D.R. (1997). Pursuing the psychopath: Capturing the fledgling psychopath in a nomological net. *Journal of Abnormal Psychology, 106,* 425–438.

Lynam, D.R., Caspi, A., Moffit, T.E., Loeber, R., & Stouthamer-Loeber, M. (2007). Longitudinal evidence that psychopathy scores in early adolescence predict adult psychopathy. *Journal of Abnormal Psychology, 116,* 155–165.

Lynam, D.R., & Widiger, T.A. (2001). Using the five-factor model to represent the DSM-IV personality disorders: An expert consensus approach. *Journal of Abnormal Psychology, 110,* 401–412.

Lyons-Ruth, K. (1996). Attachment relationships among children with aggressive behavior problems: The role of disorganized early attachment patterns. *Journal of Consulting and Clinical Psychology, 64,* 64–73.

Lyons-Ruth, K., Easterbrooks, M.A., & Davidson, C.C. (1997). Infant Attachment Strategies, Infant Mental Lag, and Maternal Depressive Symptoms: Predictors of Internalizing and Externalizing Problems at Age 7. *Developmental Psychology. 33,* 681–692.

Lyons-Ruth, K., Alpern, L., & Repacholi, B. (1993). Disorganised infant attachment classification and maternal psychosocial problems as predictors of hostile-aggressive behavior in the preschool classroom. *Child Development, 64,* 572–585.

MacBrayer, E.K., Milich, R., & Hundley, M. (2003). Attributional biases in aggressive children and their mothers. *Journal of Abnormal Psychology, 112,* 698–708.

MacDonald, A.J. (1979). Review: children and companion animals. *Child: Care, Health & Development, 5,* 347–358.

Macdonald, J.M. (1961). *The Murderer and His Victim.* Springfield, IL: Charles C. Thomas.

Macdonald, J.M. (1963). The threat to kill. *American Journal of Psychiatry, 120,* 125–130.

McCrae, R.R., & Costa, P.T. (1990). *Personality in Adulthood.* New York: Guilford Press.

McLoyd, V.C. (1990). The impact of economic hardship on Black families and children: Psychological distress, parenting, and socioemotional development. *Child Development. 61,* 311–346.

McPhedran, S. (2009). A review of the evidence for associations between empathy, violence, and animal cruelty. *Aggression and Violent Behavior, 14,* 1–4.

Main, M., & Hesse, E. (1990). Parents' unresolved traumatic experiences are related to infant disorganized attachment status: Is frightened or frightening parental behavior the linking mechanism? In M.T. Greenberg, D. Cicchetti, & E.M. Cummings (eds.), *Attachment in the Preschool Years* (pp. 161–182). Chicago: University of Chicago Press.

Marcus-Newhall, A., Pederson, W.C., Carlson, M., & Miller, N. (2000). Displaced aggression is alive and well: A meta-analytic review. *Journal of Personality and Social Psychology, 78,* 670–689.

Margolin, G., & Gordis, E.B. (2000). The effects of family and community violence on children. *Annual Review of Psychology, 51,* 445–479.

Marsee, M.A., & Frick, P.J. (2010). Callous-Unemotional traits and aggression in youth. In W.F. Arsenio & E.A. Lemerise (eds.), *Emotions, Aggression, and Morality in Children: Bridging Development and Psychopathology.* (pp. 137–156). Washington, D.C.: American Psychological Association.

Matsumoto, D, & Juang, L. (2008) *Culture and Psychology (4th ed.).* Australia: Thomson Wadsworth.

Maughan, A., & Cicchetti, D. (2002). Impact of child maltreatment and interadult violence on children's emotion regulation abilities and socioemotional adjustment. *Child Development, 73,* 1525–1542.

Mayer, S.E., & Jencks, C. (1989). Growing up in poor neighborhoods: How much does it matter? *Science. 243,* 1441–1445.

Mead, M. (1964). Cultural factors in the cause and prevention of pathological homicide. *Bulletin in the Menninger Clinic, 28,* 11–22.

Mellor, D., Yeow, J., Hapidzal, N.F.M., Yamamoto, T., Yokoyama, A., & Nobuzane, Y. (2009). *Childhood Cruelty to Animals: a Tri-National Study.*

Mellor, D., Yeow, J., bt Mamat, N. H., & bt Mohd Hapidzal, N. F. (2008). The relationship between childhood cruelty to animals and psychological adjustment: A Malaysian study. *Anthrozoos, 21,* 363–374.

Melson, G.F. (1988). Availability of and involvement with pets by children: Determinants and correlates. *Anthrozoos, 2,* 45–52.

Melson, G.F. (1990). Studying children's attachment to their pets: A conceptual and methodological review. *Anthrozoos, 4,* 91–99.

Melson, G.F. (2003). Child development and the human-companion animal bond. *American Behavioral Scientist, 47,* 31–39.

Merz-Perez, L., Heide, K.M., & Silverman, I.J. (2001). Childhood cruelty to animals and subsequent violence against humans. *International Journal of Offender Therapy and Comparative Criminology, 45,* 556–573.

Merz-Perez, L., & Heide, K.M. (2004). *Animal Cruelty: Pathway to Violence against People.* Walnut Creek, CA: AltaMira Press.

Milgram, S. (1974). Obedience to authority: An experimental view. New York: Harper.

Miller, C. (2001). Childhood animal cruelty and interpersonal violence. *Clinical Psychology Review, 21*(5), 735–749.

Miller, K.S., & Knutson, J.F. (1997). Reports of severe physical punishment and exposure to animal cruelty by inmates convicted of felonies and by university students. *Child Abuse and Neglect, 21,* 59–82.

Miller, J.D., & Lynam, D.R. (2001). Structural models of personality and their relation to antisocial behaviour: A meta-analytic review. *Criminology, 70,* 765–798.

Miller-Perrin, C.L. Perrin, R.D., Kocur, J.L. (2009). Parental physical and psychological aggression: Psychological symptoms in young adults. *Child Abuse & Neglect. 33,* 2009, 1–11.

Mischel, W. (1973). Toward a cognitive social learning reconceptualization of personality. *Psychological Review, 80,* 252–283.

Mischel, W. (1974). Processes in delay of gratification. In L. Berkowitz (ed.), *Advances in Experimental Social Psychology.* (Vol. 7, pp. 249–292). New York: Academic Press.

Moffitt, T.E. (1990). Juvenile delinquency and attention-deficit disorders: Boys' developmental trajectories from age 3 to age 15. *Child Development, 61,* 893–910.

Moffitt, T.E. (1993). Adolescence-limited and life-course-persistent antisocial behavior: A developmental taxonomy. *Psychological Review, 100,* 674–701.

Moffitt, T.E. (2003). Life-course persistent and adolescent-limited antisocial behavior: A 10-year research review and research agenda. In B.B. Lahey, T.E., Moffitt, & A. Caspi (eds.), *Causes of Conduct Disorder and Juvenile Delinquency* (pp. 49–75), New York: Guilford Press.

Moffitt, T.E. (2006). Life-course-persistent versus adolescence-limited antisocial behavior. In D. Cicchetti & D.J. Cohen (eds.), *Developmental Psychopathology, Vol. 3: Risk, Disorder, and Adaptation* (2nd ed., pp. 570–598). Hoboken, NJ: John Wiley & Sons.

Moffitt, T.E., & Caspi, A. (2001). Childhood predictors differentiate life-course persistent and adolescence-limited antisocial pathways among males and females. *Development and Psychopathology, 13,* 135–151.

Moffit, T.E., Caspi, A., Rutter, M., & Silva, P.A. (2001). *Sex Differences in Antisocial Behaviour: Conduct Disorder, Delinquency, and Violence in the Dunedin Longitudinal Study.* Cambridge, England, Cambridge University Press.

Moffitt, T.E., Krueger, R.F., Caspi, A., & Fagan, J. (2000). Partner abuse and general crime: How are they the same? How are they different? *Criminology, 38,* 199–232.

O' Connor, T.G. (2002). Annotation: The "effects" of parenting reconsidered: Findings, challenges, and applications. *Journal of Child Psychology and Psychiatry, 43,* 555–572.

Olsen, G.W., Quigley, J.S., & Stryker, R. (1988). Companion animals in the environment. In G.W. Gordon & R. Stryker (eds.), Creative long-term care administration (2nd ed., pp. 274–285). Springfield, IL, England: Charles C. Thomas.

Olweus, C.R. (1995). Bullying or peer abuse at school: Facts and intervention. *Current Directions in Psychological Science, 4,* 196–200.

Pagani, C., Robustelli, F., and Ascione, F.R. (2007). Italian youths' attitudes toward and concern for animals. *Anthrozoos, 20,* 275–293.

Parke, R.D. & Duer, J.L. (1972). Schedule of punishment and inhibition of aggression in children. *Developmental Psychology, 7,* 266–269.

Patterson, G.R., Dishion, T.J. & Bank, L. (1984). Family interaction: A process model of deviancy training. *Aggressive Behavior, 10,* 253–267.

Patterson, G.R., DeBaryshe, B.D., & Ramsey, E. (1989). A developmental perspective on antisocial behavior. *American Psychologist, 44,* 329–335.

Patterson, G.R., Littman, R.A., & Bricker, W. (1967). Assertive Behavior in Children: A Step toward a Theory of Aggression. *Monographs of the Society for Research in Child Development, 32,* 5.

Patterson, G.R., Reid, J.B., & Dishion, T.J. (1992). *A social learning approach: Vol. 4. Antisocial boys.* Eugene, OR: Castalia Press.

Pederson, W.C., Gonzales, C., & Miller, N. (2000). The moderating effect of trivial triggering provocation on displaced aggression. *Journal of Personality and Social Psychology, 78,* 913–927.

People for the Ethical Treatment of Animals (2003). *Animal abuse and human abuse: Partners in crime.* http://www.peta.org/issues/Companion-Animals/animal-abuse-and-human-abuse-partners-in-crime.aspx. Retrieved 23rd July, 2012.

Peterson, L., Gable, S., Doyle, C., & Ewugman, B. (1998). Beyond parenting skills: Battling barriers and building bonds to prevent child abuse and neglect. *Cognitive and Behavioral Practice, 4,* 53–74.

Petersen, M.L., & Farrington, D.P. (2007). Cruelty to animals and violence to people. *Victims & Offenders, 2,* 21–43.

Pollak, S.D., & Tolley-Schell, S.A. (2003). Selective Attention to Facial Emotion in Physically Abused Children. *Journal of Abnormal Psychology, 112,* 323–338.

Poresky, R.H. (1990). The Young Children's Empathy Measure: Reliability, validity and effects of companion animal bonding. *Psychological Reports, 66,* 931–936.

Poresky, R.H., Hendrix, C., Mosier, J. E., & Samuelson, M.L. (1988). Young children's companion animal bonding and adults' pet attitudes: A retrospective study. *Psychological Reports, 62*(2), 419–425.

Poresky, R.H., & Hendrix, C. (1990). Differential effects of pet presence and pet-bonding on young children. *Psychological Reports, 67*(1), 51–54.

Pulkkinen, L. (1990). Adult life-styles and their precursors in the social behaviour of children and adolescents. *European Journal of Personality, 4,* 237–251.

Pulkkinen, L. (1996). Proactive and reactive aggression in humans: A longitudinal perspective. *Aggressive Behavior, 22,* 241–257.

Quiggle, N.L., Garber, J., Panak, W.F., & Dodge, K.A. (1992). Social information processing in aggressive and depressed children. *Child Development, 63,* 1305–1320.

Quinslick, J.A. (1999). Animal abuse and family violence. In F.R. Ascione & P. Arkow (eds.), *Child Abuse, Domestic Violence, and Animal Abuse: Linking the Circles of Compassion for Prevention and Intervention.* (pp. 168–175). Indiana: Purdue University Press.

Raine, A. (2002). Biosocial studies of antisocial and violent behavior in children and adults: A review. *Journal of Abnormal Child Psychology, 30,* 311–326.

Raine, A., Brennen, P.A., Farrington, D.P., & Mednick, S.A. (eds.). (1997). *Biosocial Bases of Violence.* London: Plenum.

Raine, A., Reynolds, C., Venables, P.H., Mednick, S.A. (2002). Stimulation-seeking and intelligence: A prospective longitudinal study. *Journal of Personality and Social Psychology. 82,* 663–674.

Reebye, P. (2005). Aggression during early years: Infancy and preschool. *Canadian Child and Adolescent Psychiatry Review, 14,* 16–20.

Regan, T. (1983). *The Case for Animal Rights,* Los Angeles: University of California Press.

Renken, B., Egeland, B., Marvinney, D., Mangelsdorf, S., & Sroufe, L.A. (1989). Early childhood antecedents of aggression and passive-withdrawal in early elementary school. *Journal of Personality, 57,* 257–281.

Repetti, R.L., Taylor, S.E., & Seeman, T.E. (2002). Risky families: Family social environments and the mental and physical health of offspring. *Psychological Bulletin, 128,* 330–366.

Ressler, R.K., Burgess, A.W., Hartman, C.R., Douglas, J. E., & McCormack, A. (1986). Murderers who rape and mutilate. *Journal of Interpersonal Violence, 1,* 273–287.

Ressler, R.K., Burgess, A.W., & Douglas, J.E. (1988). *Sexual Homicide: Patterns and Motives.* Lexington, Mass: Lexington Books.

Rhee, S.H., & Waldman, I.D. (2011). Genetic and environmental influences on aggression. In P.R. Shaver & M. Mikulincer, M. (eds.), *Human Aggression and Violence: Causes, Manifestations, and Consequences.* (pp. 143–163). Washington, D.C.: American Psychological Association.

Rieder, C., & Cicchetti, D. (1989). Organizational perspective on cognitive control functioning and cognitive affective balance in maltreated children. *Developmental Psychology, 25,* 382–393.

Rigby, K. (2002). Bullying in childhood. In P.K. Smith, & C.H. Craig (eds.), *Blackwell Handbook of Childhood Social Development.* (pp. 549–568). Malden: Blackwell Publishing.

Rigdon, J.D., & Tapia, F. (1977). Children who are cruel to animals: A follow-up study. *Journal of Operational Psychiatry, 8,* 27–36.

Robins, L.N. (1991). Conduct Disorder. *Journal of Child Psychology and Psychiatry, 32.* 193–212.

Rogeness, G.A., Cepeda, C., Macedo, C.A., Fischer, C., & Harris, W.R. (1990). Differences in heart rate and blood pressure in children with conduct disorder, major depression, and separation anxiety. *Psychiatry Research. 33,* 199–206.

Rollins, B.E. (2006). *Animal Rights & Human Dignity (3rd ed.).* New York: Prometheus Books.

Rothbart, M.K & Bates, J.E. (1998). Temperament. In N. Eisenberg (ed.). (1998). *Handbook of Child Psychology,* (5th ed.): *Vol. 3. Social, Emotional, and Personality Development.* (pp. 105–176). Hoboken, NJ: John Wiley & Sons.

Rothbaum, F., & Weisz, J. (1994). Parental caregiving and child externalizing behavior in nonclinical samples: A meta-analysis. *Psychological Bulletin, 116,* 55–74.

Rule, B.G., & Ferguson, T.J. (1986). The effects of media violence on attitudes, emotions, and cognitions. *Journal of Social Issues, 42,* 29–50.

Rutter, M. (1989). Pathways from childhood to adult life. *Journal of Child Psychology and Psychiatry, 30,* 25–31.

Rutter, M. (2003). Commentary: Causal processes leading to antisocial behavior. *Developmental Psychology, 39,* 372–378.

Rutter, M., Cox, A., Tupling, C., Berger, M., & Yule, W. (1975). Attainment and adjustment in two geographical areas: I. The prevalence of psychiatric disorder. *British Journal of Psychiatry, 126,* 493–509.

Rutter, M., Tizard, J., & Whitmore, K. (eds.) (1970). *Education, Health and Behaviour.* London: Longman.

Ryder, R.D. (1973). Pets in man's search for sanity. *Journal of Small Animal Practice, 14,* 657–668.

Sampson, R.J., & Laub, J.H. (1990). Crime and deviance over the life course: The salience of adult social bonds. *American Sociological Review, 55*, 609–627.

Sampson, R.J. & Laub, J.H. (1994). Urban poverty and the family context of delinquency: A new look at structure and process in a classic study. *Child Development. 65*, 523–540.

Sampson, R.J., Raudenbush, S. W., & Earls, F. (1997). Neighborhoods and violent crime: A multilevel study of collective efficacy. *Science, 277*, 918–924.

Schaffner, J.E. (2009). Laws and policy to address the link of family violence. In A. Lindzey (ed.), *The Link between Animal Abuse and Human Violence* (pp. 228–237). Sussex: Academic Press.

Schlueter, S. (2008). Law enforcement perspectives and obligations related to animal abuse. In F. Ascione (ed.), *The International Handbook of Theory and Research on Animal Abuse and Cruelty* (pp. 375–391). West Lafayette, Indiana: Purdue University Press.

Schneider, W., & Shiffrin, R.M. (1977). Controlled and automatic human information processing: I. Detection, search and attention. *Psychological Review, 84*, 1–66.

Schultz, D., Izard, C.E., & Bear, G. (2004). Children's emotion processing: Relations to emotionality and aggression. *Development and Psychopathology, 16* 371–387.

Schwartz, C.E., Wright, C.I., Shin, L.M., Kagan, J., & Rauch, S.L. (2003). Inhibited and uninhibited infants "Grown Up": Adult amygdalar response to novelty. *Science, 300, no. 5627*, 1952–1953.

Schwartz, D., McFayden-Ketchum, S.A., Dodge, K.A., Pettit, G.S., & Bates, J.E. (1998). Peer group victimization as a predictor of children's behavior problems at home and in school. *Development and Psychopathology, 10*, 87–99.

Seligman, M.E.P. (1970). On the generality of the laws of learning. *Psychological Review, 77*, 406–418.

Selman, R.L. (1980). *The Growth of Interpersonal Understanding: Developmental and Clinical Analyses.* New York: Academic Press.

Serbin, L.A., & Karp, J. (2004). The intergenerational transfer of psychosocial risk: Mediators of vulnerability and resilience. *Annual Review of Psychology, 55*, 333–363.

Shantz, D.W., & Voydanoff, D.A. (1973). Situational effects on retaliatory aggression at three age levels. *Child Development, 44*, 149–153.

Shields, A., & Cicchetti, D. (1998). Reactive aggression among maltreated children: The contributions of attention and emotion dysregulation. *Journal of Clinical Child Psychology, 27*, 381–395.

Simmons, C. A., & Lehmann, P. (2007). Exploring the link between pet abuse and controlling behaviors in violent relationships. *Journal of Interpersonal Violence, 22*, 1211–1222.

Simons, K. J., Paternite, C. E., & Shore, C. (2001). Quality of parent/adolescent attachment and aggression in young adolescents. *Journal of Early Adolescence, 21*, 182–203.

Simons, R.L., Wu, C.I., Conger, R.D., & Lorenz, F.O. (1994). Two routes to delinquency: Differences between early and late starters in the impact of parenting and deviant peers. *Criminology, 32*, 247–276,

Slavkin, M.L. (2001). Enuresis, firesetting, and cruelty to animals: Does the ego triad show predictive validity? *Adolescence, 36*, 461–466.

Smith, P.K., & Myron-Wilson, R. (1998). Parenting and school bullying. *Clinical Child Psychology and Psychiatry, 3,* 405–417.

Snyder, J., & Patterson, G. R. (1995). Individual differences in social aggression: A test of a reinforcement model of socialisation in the natural environment. *Behavior Therapy, 26,* 371–391.

Spergel, I.A., & Curry, G.D. (1993). The National Youth Gang Survey: A research and development process. In A.P. Goldstein, & R.C. Huff (eds.) *The Gang Intervention Handbook.* (pp. 359–400). Champaign, IL: Research Press.

Stanger, C., Achenbach, T.M., Verhulst, F.C. (1997). Accelerated longitudinal comparisons of aggressive versus delinquent syndromes. *Development and Psychopathology. 9,* 43–58.

Staub, E. (1989). *The Roots of Evil: the Origins of Genocide and Other Group Violence.* New York: Cambridge University Press.

Staub, E. (1998). Breaking the cycle of genocidal violence: Healing and reconciliation. In J. Harvey (ed.), *Perspectives on Loss: a Sourcebook.* (pp. 23–38). Philadelphia: Taylor & Francis.

Straus, M.A. (1991). Discipline and deviance: Physical punishment of children and violence and other crime in adulthood. *Social Problems. 38,* 133–154.

Sternberg, C., & Campos, J. (1990). The development of anger expression in infancy. In N. Stein, T. Trabasso, & B. Leventhal (eds.), *Concepts in Emotion.* (pp. 75–99). Hillsdale, N.J.: Erlbaum.

Stevens, D., Charman, T., & Blair, R.J.R. (2001). Recognition of emotion in facial expressions and vocal tones in children with psychopathic tendencies. *The Journal of Genetic Psychology, 16,* 201–211.

Stifter, C.A., Spinrad, T.L., & Baungart-Rieker, J.M. (1999). Toward a developmental model of child compliance: The role of emotion regulation in infancy. *Child Development, 70,* 21–32.

Stormshak, E.A., Bierman, K. L., Bruschi, C., Dodge, K.A., Coie, J.D., & the Conduct Problems Prevention Research, G. (1999). The relation between behavior problems and peer preference in different classroom contexts. *Child Development, 70,* 169–182.

Stormshak, E.A., Bierman, K.L., McMahon, R.J., Lengua, L.J., & the Conduct Problems Prevention Research, G. (2000). Parenting practices and child disruptive behavior problems in early elementary school. *Journal of Clinical Child Psychology, 29,* 17–29.

Straus, M.A. (1991). Discipline and deviance: Physical punishment of children and violence and other crimes in adulthood. *Social Problems, 38,* 133–154.

Stromquist, V.J., & Strauman, T.J. (1991). Children's social constructs: Nature, assessment, and association with adaptive versus maladaptive behavior. *Social Cognition. 9,* 330–358.

Tackett, J.L., & Krueger, R.F. (2011). Dispositional influences on human aggression. In P.R. Shaver & M. Mikulincer, M. (eds.), *Human Aggression and Violence: Causes, Manifestations, and Consequences.* (pp. 89–104). Washington, D.C.: American Psychological Association.

Tallichet, S.E., & Hensley, C. (2004). Exploring the link between recurrent acts of childhood and adolescent animal cruelty and subsequent violent crime. *Criminal Justice Review, 29,* 304–316.

Tapia, F. (1971). Children who are cruel to animals. *Child Psychiatry & Human Development, 2,* 70–77.

Tellegen, A. (1985). Structures of mood and personality and their relevance to assessing anxiety with an emphasis on self-report. In A. Hussain Tuma and Jack D. Maser (eds.). *Anxiety and the Anxiety Disorders.* (pp. 681–706). Hillsdale, N.J.: Lawrence Erlbaum Associates.

Thomas, A., & Chess, S. (1977). *Temperament and Development.* New York: Brunner/Mazel.

Thompson, K.L., & Gullone, E. (2006). An investigation into the association between the witnessing of animal abuse and adolescents' behavior toward animals. *Society and Animals, 14,* 223–243.

Thompson, K., & Gullone, E. (2008). Prosocial and antisocial behaviours in adolescents: An investigation into associations with attachment and empathy. *Anthrozoos, 21,* 123–137.

Thompson, R.A. (1994). Emotion regulation: A theme in search of definition. *Monographs of the Society for Research in Child Development, 59,* 250–283.

Thornberry, T.P., Krohn, M.D., Lizotte, A.J., Chard-Wierschem, D. (1993). The role of juvenile gangs in facilitating delinquent behavior. *Journal of Research in Crime and Delinquency. 30*55–87.

Tingle, D., Barnard, G.W., Robbins, L., Newman, G., & Hutchinson, D. (1986). Childhood and adolescent characteristics of pedophiles and rapists. *International Journal of Law and Psychiatry, 9,* 103–116.

Todorov, A., & Bargh, J.A. (2002). Automatic sources of aggression. *Aggression and Violent Behavior, 7,* 53–68.

Torgersen, S., Kringlen, E., & Cramer, V. (2001). The prevalence of personality disorders in a community sample. *Archives of General Psychiatry, 58,* 590–596.

Toth, S.L., Manly, J.T., & Cicchetti, D. (1992). Child maltreatment and vulnerability to depression. *Developmental Psychopathology, 4,* 97–112.

Tremblay, R.E. (2000). The development of aggressive behaviour during childhood: What have we learned in the past century? *International Journal of Behavioral Development, 24,* 129–141.

Tremblay, R.E. Loeber, R., Gagnon, C., Charlebois, P., Larivee, S., & LeBlanc, M. (1991). Disruptive boys with stable and unstable high fighting behavior patterns during junior and elementary school. *Journal of Abnormal Psychology, 19,* 285–300.

Troy, M. & Sroufe, L.A. (1987) Victimization among preschoolers Role of attachment relationship history *Journal of the American Academy of Child and Adolescent Psychiatry, 26,* 166–172.

Underwood, M.K. (2003). *Social Aggression Among Girls.* New York: Guilford Press.

United Nations (2011). http://www.un.org/en/ Retrieved 2 August 2010.

Unti, B. (2008). Cruelty indivisible: Historical perspectives on the links between cruelty to animals and interpersonal violence. In F.R. Ascione (ed.). *The International Handbook of Animal Abuse and Cruelty: Theory, Research, and Application* (pp. 7–30). West Lafayette, Indiana: Purdue University Press.

Urbina, I. (2010). Animal abuse as clue to additional cruelties. *The New York Times reprints.* 17th March, 2010. http://www.nytimes.com/2010/03/18/us/18animal.html retrieved on 12 April 2011.

Vaughn, M.G., Fu, Q., DeLisi, M., Beaver, K.M., Perron, B.E., Terrell, K., et al. (2009). Correlates of cruelty to animals in the United States: Results from the National Epidemiologic Survey on Alcohol and Related Conditions. *Journal of Psychiatric Research, 43,* 1213–1218.

Veenstra, R., Lindenberg, S., Oldehinkel, A.J., De Winter, A.F., Verhulst, F.C., & Ormel, J. (2005). Bullying and victimization in elementary schools: A comparison of bullies, victims, bully/victims, and uninvolved preadolescents. *Developmental Psychology, 41*, 672–682.

Verlinden, S., Hersen, M., & Thomas, J. (2000). Risk factors in school shootings. *Clinical Psychology Review, 20*, 3–56.

Vidovic, V.V., Stetic, V.V. and Bratko, D. (1999). Pet ownership, type of pet and socio-emotional development of school children. *Anthrozoos, 12*, 211–217.

Vitaro, F., Brendgen, M., Pagani, L., Tremblay, R.E., & McDuff, P. (1999). Disruptive behavior, peer association, and conduct disorder: Testing the developmental links through early intervention. *Development and Psychopathology. 11*, 287–304.

Vitaro, F., Brendgen, M., & Tremblay, R.E. (2002). Reactively and proactively aggressive children: Antecedent and subsequent characteristics. *Journal of Child Psychology and Psychiatry, 43*, 495–506.

Volant, A.M., Johnson, J.A., Gullone, E., & Coleman, G.J. (2008). The relationship between domestic violence and animal abuse: An Australian study. *Journal of Interpersonal Violence, 23*(9), 1277–1295.

Walter, H.J. (2001). School-based prevention of problem behaviors. *Child & Adolescent Psychiatric Clinics of North America, 10*, 117–127.

Walton-Moss, B.J., Manganello, J., Frye, V., & Campbell, J.C. (2005). Risk Factors for Intimate Partner Violence and Associated Injury Among Urban Women. *Journal of Community Health: The Publication for Health Promotion and Disease Prevention, 30*, 377–389.

Wax, D., & Haddox, V. (1974a). Enuresis, firesetting, and animal cruelty in male adolescent delinquents: A triad predictive of violent behavior. *Journal of Psychiatry & Law, 2*, 45–71.

Wax, D.E., & Haddox, V.G. (1974b). Enuresis, fire setting, and animal cruelty: A useful danger signal in predicting vulnerability of adolescent males to assaultive behavior. *Child Psychiatry & Human Development, 4*, 151–156.

Weinberg, K.M., Tronick, E.Z., Cohn, J.F., & Olson, K.L. (1999). Gender differences in emotional expressivity and self-regulation during early infancy. *Developmental Psychology, 35*, 175–188.

Widom, C.S. (1989). Does violence beget violence? A critical examination of the literature. *Psychological Bulletin, 106*, 3–28.

Wolfe, D.A., & Mosk, M.D. (1983). Behavioral comparisons of children from abusive and distressed families. *Journal of Consulting and Clinical Psychology, 51*, 702–708.

Worsham, N. L., & Goodvin, R. (2007). The Bee Kind Garden: A qualitative description of work with maltreated children. *Clinical Child Psychology and Psychiatry, 12*, 261–279.

Wright, J.C., Giammarino, M., & Parad, H.W. (1986). Social status in small groups: Individual-group similarity and the social "misfit". *Journal of Personality and Social Psychology, 50*, 523–536.

Wright, J., & Hensley, C. (2003). From animal cruelty to serial murder: Applying the graduation hypothesis. *International Journal of Offender Therapy and Comparative Criminology, 47*, 71–88.

Zahn-Waxler, C., Radke-Yarrow, M., & King, R.A. (1979). Child rearing and children's prosocial initiations toward victims of distress. *Child Development, 50*(2), 319–330.

Zigler, E., Taussig, C., Black, K. (1992). Early childhood intervention: A promising preventative for juvenile delinquency. *American Psychologist, 47*, 997–1006.

Glossary

Aetiology: Theories or pathways that describe or hypothesise how individual characteristics, behaviours, personality traits or particular symptoms develop.

Affect: usually refers to emotion. See also "Processes of development" entry below.

Affection: A term for a disposition or state of mind (or body) that describes a feeling of caring or type of love.

Age of onset: The age at which an individual acquires, develops, or first experiences a condition or symptoms of a disease or disorder.

Antecedent: A preceding event, condition, cause, phrase, or word.

Anxiety: A psychological and physiological state proposed to comprise somatic, emotional, cognitive, and behavioural components. Anxiety is classified as a negative, but normal (not pathological), emotion. It can, however, be considered pathological if it manifests in such a way that it interferes with everyday functioning.

Attachment: Describes a type of relationship that is characterised by a strong affectionate bond. Such a relationship tends to develop with others who have special significance in our lives. For children, the strongest attachment bond usually develops with the primary caregiver.

Attachment Styles

Secure Attachment: This is the psychologically healthiest attachment style. Infants who learn this style develop confidence in their ability to regulate emotion, including containing impulses when necessary, expressing feelings when appropriate, and becoming emotionally invested in activity. In later development, negative emotions are not experienced as threatening but are seen as providing a communicative function. Children who develop secure attachment develop healthy emotion regulation. They develop trust in relationships and are comfortable forming close bonds.

Avoidant-resistant Attachment: Infants in this category learn an overly rigid style of emotion regulation as a result of experiencing a caregiver who repeatedly ignores the infant or actively rejects the child's expressions of distress or attempts to gain reassurance. In later development, given their extremely distressing emotional experiences when faced with threat, these children's attentional processes are likely to be modified, and they are likely to avoid emotionally arousing situations. Since these infants have internalized beliefs that only a restricted range of emotions is acceptable, their self-regulation of emotion will likely be characterized by restrictions or distortions in experience and expression.

Ambivalent or Anxious Attachment: Children who develop this attachment style have typically experienced unpredictable and intermittent caregiver responsiveness to their distress signals. Their communications of distress or anger do not predictably lead to emotion restabilisation or to a sense of security in their caregiver's presence. These infants therefore remain vigilant and are likely to heighten their expressions of distress in an effort to gain their caregiver's attention. Their experiences of being ineffective at regulating emotion leads to a sense of insecurity about separating from the caregiver and may support a self-regulatory style characterized by heightened arousal, exaggerated emotional expression, and a view of self as unworthy and/or incompetent.

Anxious-disorganized Attachment: This insecure attachment style develops when the attachment figure is in some way frightened of, or frightening to, the child. Maltreated infants and infants of psychotic parents tend to develop this attachment style.

Attention deficits: Attentional problems; for example, limited ability to concentrate or focus on the task at hand.

Attribution: A concept that refers to the way in which individuals understand or explain causes of behaviour or events.

Authoritarian parenting: Parenting that is characterised by extremely restrictive and controlling parenting behaviours. Authoritarian parents expect rules they have set to be followed without argument. Any disobedience of rules is punished, and debate of rules is firmly discouraged. Power assertion is heavily relied upon, and compliance is demanded. Coercive techniques such as threats or physical punishment are relied upon rather than reasoning or explanation.

Automatized: Behaviours or thoughts that become automatic over time. They therefore largely lose their intentional and volitional nature.

Baseline level: A base for measurement or a point of reference.

Behavioural scripts: A term used to describe a sequence of expected behaviours for a given situation. We continually follow scripts which we acquire over the course of development and/or time. For example, when we attend a church service, we understand that we are expected to remain silent and relatively still.

Biological: See "Processes of development" entry below.

Biological preparedness: Seligman (1971) argued that certain stimuli are biologically significant. Evolution has predisposed organisms to easily acquire associations that facilitate survival of the species. The organism is therefore prepared to learn to respond to certain biologically significant stimuli that may have threatened or promoted survival in the species' evolutionary past.

Callous attitudes: A response style that is reward-oriented and low in empathy. Callous attitudes are characterised by a lack of reactivity to signs of distress in others.

Causal connection: A connection between a stimulus and a response. For example, there is a causal connection between experiencing an event or communication as funny and responding with laughter.

CBCL: Acronym for the Child Behavior Checklist – a commonly used measure that assesses children and adolescent's functioning on both internalising and externalising dimensions.

Coercion: The use of force or pressure causing another party to behave in an involuntary manner (whether through action or inaction).

Cognition: Mental processes. Such processes include attention, memory, language comprehension and production, solving problems, and decision-making. See also "Processes of development" entry below.

Conduct Disorder: "A repetitive and persistent pattern of behaviour in which the basic rights of others or major age-appropriate societal norms or rules are violated" (Diagnostic and Statistical Manual-IV – Text Revised (DSM-IV-TR; American Psychiatric Association, 2000, p. 98).

Conscience: An aptitude, faculty, intuition, or judgment of the intellect that distinguishes right from wrong.

Constitutional: Aspects of a person that are believed to be inherent. It particularly refers to morphological or physiological aspects that are organic, genetic, and relatively stable over the course of development.

Contextual: Events, processes or environments that characterise a particular situation believed to have an impact on a person's behaviour.

Continuity: A characteristic relating to an individual or behaviour such that it displays continuity over time and across developmental periods.

Co-occurrence: The overlapping occurrence of two or more behaviours, events, or phenomena.

Cross-reporting: The reporting of an event, situation, behaviour, or other phenomena by more than one person or party. For example, the reporting of animal cruelty by an animal welfare officer and a parent.

Dehumanization: Describes efforts to mitigate one's sense of humanity or to undermine one's access to basic human rights. It is often accomplished either through language (e.g., verbal abuse), symbolically (e.g., imagery), or physically (e.g., physical abuse, refusing eye contact).

Delaying of gratification: The ability to delay or wait for rewards to achieve a more important goal. For example, a student is delaying gratification if they delay recreational outings until after the completion of school exams.

Deviance: Thoughts, actions or behaviours that violate social norms or expectations and, at an extreme level include behaviours that violate societal laws (e.g., crime).

Diagnostic and Statistical Manual of Mental Disorders (DSM): The most recent version of this manual is the text revised fourth edition (DSM-IV-*TR*): It is a manual used for the classification of mental disorders. It was published by the American Psychiatric Association with the goal of providing a common language and standard criteria for the classification of mental disorders.

Displaced aggression: Constitutes a form of aggression against others (human or non-human animal) who did not play a direct role in the precipitating event. Displaced aggression increases if the target of such aggression provides even a minor trigger or the slightest of provocations (e.g., a dog barking). Displaced aggression also increases if the target can be perceived to be a member of a disliked out-group or as having less social value (e.g., a non-human animal).

Dispositional: A habit or a tendency to act in a specified way.

Drive theory: This theory is based on the belief that organisms are born with certain psychological needs and that tension results when these needs are not satisfied. If the need is satisfied, the drive is reduced, and the organism returns to a state of homeostasis (i.e., a stable and constant state).

Dysfunctional family: A Family that is characterised by conflict. Such families are often abusive.

Ecological factors: Environmental factors that interact in important ways with the functioning and adaptation of living beings, including humans and animals.

Effect sizes: A statistical term that represents a measure of the strength of the relationship between two or more variables.

Effortful control: Involves a number of key abilities including voluntarily focusing and shifting attention, as well as the abilities to inhibit or initiate behaviour consistent with the demands of the situation.

Egocentrism: Characterised by preoccupation with one's own internal world to such an extent that perceptions of one's environment are biased toward one's own world view and important aspects of the situation may be ignored or overlooked.

Electroencephalogram: A device used to record electrical activity along the scalp. The Electroencephalogram or EEG measures voltage fluctuations that result from ionic current flows within the neurons of the brain.

Emotion: A complex psychophysiological experience of an individual's state of mind. Emotions are believed to result from an interaction between biochemical (internal) and environmental (external) factors. Emotions are associated with moods, temperament, personality, disposition, and motivation.

Emotion dysregulation: Emotional functioning that does not fall within the conventionally accepted range of emotive response. Examples of emotion dysregulation include angry outbursts or aggressive behaviours, such as destroying or throwing objects. Emotional dysregulation is often present in people with psychiatric or psychological disorders.

Emotion processes: Refer to the "Processes of development" entry below.

Empathic arousal: Describes an individual's affective response that relates to another's emotional state. With empathy, the arousal or response experienced is perceived by the individual as being similar to that experienced by the other.

Empirically supported: A term used within Evidence-based Practice (EBP) which is an interdisciplinary approach based on the principles that sound practical decisions should follow research or empirical studies that meet a specific set of criteria. Such criteria point to methodologically sound research that enables valid outcomes and conclusions.

Encoding: A process that allows information to be stored and recalled in memory for immediate or later use or manipulation.

Enuresis: A repeated inability to control urination. Use of the term is usually limited to describing individuals old enough to be expected to exercise such control.

Euphemistic language: The substitution of mild, vague, or less offensive language in place of harsh, blunt, or direct language.

Experiential factors: Factors that relate to direct experience.

Externalizing behaviours/spectrum: A spectrum that encompasses personality traits that are disinhibitory in nature, such as impulsivity and aggression. Individuals at risk of developing externalizing disorders have a temperamental style that is low in inhibition and low in control.

Flight response: The fight-or-flight response refers to the alarm or fear response upon the perception of danger. In such situations, the impulse is to flee (flight) or to stay and defend oneself (fight).

Gender: Socially influenced characteristics including gender stereotypes (widely held beliefs about characteristics deemed appropriate for males and females).

Generalizability: The application of a conclusion or finding from one group or sample to another.

Genetic: Gene behavior in relation to a cell or an organism. It can also refer to patterns of inheritance from parent to offspring.

Grandiosity: Describes an individual's sense of superiority, uniqueness, and the belief that few others have anything in common with oneself. People with a grandiose view of themselves also believe that can be understood only by a few people or by very special people.

Habituation: A decrease in an elicited behaviour or response (cognitive or emotional) that results from repeated exposure to a particular stimulus.

Heritability: Analyses the relative contribution made by genetic and non-genetic factors to the total phenotypic variance of a particular construct in a population.

Heterogeneity: Generally refers to dissimilarity and may refer to a construct being composed of dissimilar behaviours or parts.

Hostile perception bias: A tendency by some individuals who are more aggression-prone to be more likely to perceive hostility even where there is no hostility, compared to someone who is not aggression-prone.

Hostility: An emotional state that is characterised by hatred or enmity to others and that involves the desire to harm those others or to cause them pain.

Hyperactivity: A state in which there is inappropriate or excessive motor activity.

Impulsivity: The inclination or tendency to initiate behaviour without adequate reflection or forethought regarding the consequences of their behaviour.

Intergenerational: Passed from one generation to the next.

Internal working models: Cognitive representations that infants develop for their relationship with their caregiver. These later generalise to other relationships in their lives and include expectations of the availability of attachment figures.

Internalizing spectrum: This spectrum encompasses personality traits that are inhibitory in nature, such as fear and anxiety. Individuals at risk of developing internalizing disorders have a temperamental style that is high in inhibition and high in control.

Longitudinal: A research study design that involves repeated observations or measurement of the same variables with the same people (or sample) over long periods of time. Such studies are often used to study developmental trends across the life span.

Macro-: Large or more global (environments or influences such as suburb or neighbourhood) as compared to micro- or more immediate and smaller (environments or influences such as family or friend).

Micro-: Small or more immediate (environments or influences such as parents or families) as compared to macro- (see above).

Narcissism: Characteristic features include an exaggerated sense of self-importance; an excessive need for the attention and admiration of others; a tendency to overvalue one's own accomplishments and achievements; a preoccupation with fantasies about the ideal love, success, wealth or power; and inappropriate emotional reactions to being criticised by others.

Negative affect: Refers to negative emotions including distress, guilt, fear, hostility, irritability, shame, nervousness, and upset.

Normative: Pertains to norms or standards. Data collection for the purposes of determining the underlying distribution of a particular characteristic or set of characteristics in a population.

Observational learning: Also referred to as **vicarious learning**, **social learning**, or **modelling**. It is a type of learning that occurs through observation.

Outcome efficacy: The belief that certain acts will lead to desired outcomes.

Pathology: A condition or biological state in which what is considered to be normal functioning is prevented. Pathological behaviour is behaviour which is seen to beproblematic or fails to meet the requirements of adaptation in a particular society or culture and as a consequence is generally not socially accepted.

Perceptual schemas: Schemas otherwise referred to as "knowledge structures" that are stored in memory. These structures influence information processing and consequently guide our perceptions and behaviour.

Personality traits: Habitual patterns of behaving, thinking, and feeling. Personality traits are relatively stable over time, differ across individual, and are otherwise referred to as "individuals differences." For example, some people are more outgoing, whereas others are more reserved. Traits also influence people's behaviours. For example, outgoing individuals tend to prefer to spend time in the company of others, whereas less-outgoing people prefer to spend more time alone.

Predisposition: A factor or set of factors that increases the likelihood of the individual who possesses it (them) displaying particular traits or characteristics. For example, being male is a predispositional risk factor for the development of aggression.

Prevalence: Defined as the total number of cases of a specific factor in the population at a given time. It can also refer to the total number of cases in the population, divided by the number of individuals in the population. For example, prevalence can be used to provide an estimate of how common a disease or disorder is within a population over a certain period of time.

Preventative strategies: Measures taken to prevent the development of a disorder (or disorders) rather than treating them after they have already developed.

Processes of development: The study of development incorporates three separate but inter-related processes. The first of these is **biological processes**, including genes inherited from both parents at conception. A second process involves **cognitive** or thought processes such as problem-solving, logical reasoning, and creativity. The third major process involved in human development is what is referred to as the **socioemotional** process (e.g., emotion regulation, attachment relationships). This last process includes emotions and human relationships. All three processes are central to human development and are intricately and complexly related.

Prospective: Means "looking forward" and is used in psychology to refer to a methodological research design. A prospective research design looks into the future by measuring the same variables with the same people (or sample) over long periods of time as is done with longitudinal studies.

Psychopathology: Has been defined as: "A clinically significant behavioral or psychological syndrome or pattern that occurs in an individual and that is associated with present distress (e.g., a painful symptom) or disability (i.e., impairment in one or more important areas of functioning) or with a significantly increased risk of suffering death, pain, disability, or an important loss of freedom. In addition, this syndrome or pattern must not be merely an expectable and culturally sanctioned response to a particular event, for example, the death of a loved one. Whatever its original cause, it must be considered a manifestation of a behavioral, or biological dysfunction in the individual. Neither deviant behavior (e.g. political, religious, or sexual) nor conflicts that are primarily between the individual and society are mental disorders unless the deviance or conflict is a symptom of a dysfunction in the individual, as described above." (American Psychiatric Association, 1994; pp. xxi–xxii).

Retrospective: Refers to looking back at events that already have taken place. In psychology, the retrospective method involves asking participants to recall and report on past events, feelings, or experiences.

Schema: A mental set or representation.

Socialization: A process that begins at birth and involves the acquisition, through culture, of societal norms, customs, values, and ideologies. Socialization provides individuals with the skills and habits they need for adaptive participation within their own society.

Social Learning Theory: Proposes that people's learning takes place within a social context.

Societal norms: Constitute the laws that govern society's behaviours. Norms are not necessarily enshrined in laws within society. Nevertheless, they promote a great deal of social control. People who do not conform to the norms of a society are labelled as deviants.

Trajectory: The course or pathway of development of a particular characteristic or set of characteristics over time.

Unconscious mind: First put forth by Sigmund Freud, the unconscious mind is that aspect of the mind that operates outside the attention and awareness of the conscious mind.

Vicarious reinforcement: Reinforcement that is not directly experienced but observed, such as in seeing an individual being rewarded for a particular behaviour. Vicarious reinforcement can also be acquired through the information pathway as in reading or hearing about it.

Index

Abnormal, 45, 51, 57, 77, 92
Adaptive (function), 18, 27
Adolescence, 17, 25, 29–31, 34, 36, 37, 43, 46, 48, 49, 55, 62, 63, 71, 74, 75, 77, 80, 92, 93, 106, 115, 116, 121
Adolescent limited, 37, 46
Adolescent-onset, 29, 30–32, 55, 96
Adolescent period, 30, 33
Adolescent rebellion, 31
Aetiology (aetiological), 9, 33, 43, 44, 53, 91, 97, 119
Affect, 14, 21, 23, 24, 39, 40, 62, 68, 72, 81, 122
Affection, 67
Affective, 21, 22, 42, 50, 53, 64, 66, 79, 81
Affective domain, 53
Affective functioning, 50
Age of onset, 26, 35, 46, 65
Aggravated assault, 21, 30, 46
Agreeableness, 50–53, 57
Aggression
   affective, 21
   direct, 21
   hostile, 14, 21–24, 28
   impulsive, 21–23
   indirect, 20, 21, 27
   premeditated, 21–23, 26, 106
   proactive, 21–24, 26, 55
   reactive, 21–24, 26, 55
Aggressive
   criminals, 92, 94, 120
   cues, 60, 82, 124
   model, 84, 123
   parental models, 120
Aggressor, 20, 21, 111
Alcoholism, 95, 120, 122
Ambiguous situations, 22, 66, 83, 124
Anger, 14, 21, 22, 27, 28, 37, 40, 48, 49, 52, 56, 63, 65, 68, 72, 73, 77, 81, 84, 107
Anger expression, 27

Antecedent, 18, 25, 27, 34, 114
Animal
   abuse, 9, 109, 127, 136
   cruelty legislation, 132, 135
   suffering, 2, 3, 5, 11–15, 22, 125, 128, 135
   welfare societies, 93, 103, 132, 133
Antisocial behaviour
   covert, 28
   overt, 13, 63, 68, 76, 122
Antisocial Personality Disorder, 10, 26, 29, 34–36, 38, 45, 51, 53, 54, 57, 71, 94, 98, 99, 101, 102, 113, 115, 128
Anxiety, 33, 52–57, 77
Anxious-avoidant attachment, 64, 65, 66, 165
Appropriate, 4, 10–12, 26, 32, 45, 64, 79, 95, 102, 115, 124, 126, 135–137
Aquinas, Thomas, 5
Argumentative (behaviour), 26, 29, 75
Assaults, 12, 17, 21, 30, 65, 46, 96, 98, 99, 101, 102, 104, 105, 108, 121
Assertiveness, 53, 57
Attachment
   behavioural system, 43, 59, 64
   relationships, 63–67, 73, 79, 84, 131
Attention (need for), 3, 27, 128
Attention deficits, 31
Attention-shifting, 28
Attitude formation, 119
Attitudes, 2, 3, 25, 32, 74, 75, 77, 78, 83, 88, 89, 95, 114, 119, 124, 125, 128, 131, 132, 135, 138
   family, 32
   parental, 70
Attribution, 78, 80, 82–84, 87, 89, 124
Authoritarian, 68
Automatic, 22, 23, 60, 81
Automatized, 41, 81
Aversive, 40, 52, 69, 85
Avoidant-resistant attachment, 64

Baseline, 41, 46, 47, 166
Batterer, 110
Behavioural
    domain/dimension, 2, 12, 53
    problems, 10
    scripts, 81
    style, 32, 49, 54
    tendencies, 39, 80
Beliefs, 13, 42, 51, 52, 65, 75, 78, 81,
    82, 84, 85, 89, 90, 95, 107, 118,
    119, 124
Biological, 23, 40, 43–45, 47, 48,
    56–58, 114, 122
Blood pressure, 47
Bond, 64–66, 76, 79, 126
Bullying, 18, 20, 96, 98, 99, 111–115,
    117, 118, 120, 128

Callous-Unemotional traits, 14, 32,
    33, 43, 50, 83, 51, 54–57, 96, 97,
    100, 109, 124, 126, 127
Cascade of risk, 63
Causal connection, 61, 167
CBCL, 10, 167
Chaotic home environment, 120
Child abuse, 34, 71, 86, 94, 113, 120,
    121, 128, 129, 136
Child Behavior Checklist, 10
Childhood, 16–18, 24, 25, 27, 29–31,
    35–38, 42–43, 45, 48, 49, 55–57,
    61–63, 65, 66, 70, 73–76, 80,
    93–95
    animal cruelty, 3, 13
    history, 91, 92
Child-onset, 28, 31–33, 55, 96
Chronic, 1, 17, 24, 26, 53, 54, 60, 71,
    84
Cloninger's Temperament and
    Character Model, 50–52
Coercion, 67, 167
Cognition, 39, 62, 78, 89, 125
Cognitive, 3, 25, 28, 39–44, 48, 56,
    57, 60, 78, 80, 82, 85, 86, 88, 89,
    116, 124–125, 128
Cold, 21, 63, 122
Cold hearted, 53
Collective violence, 30
Community, 3, 30, 32, 35, 43, 59,
    100, 109, 110, 115, 134, 135

Community connectedness, 32
Comorbidity, 36, 109
Compassion, 4–6, 52, 125
Competencies, 10, 32, 33, 68, 78, 79,
    126
Conduct disorder, 3, 9, 15, 18, 26,
    29, 34–38, 46, 47, 67, 95–102,
    113–115, 126, 128, 132, 134
Conflict(s), 3, 27, 31, 32, 43, 63, 67,
    76, 112, 122
Conflictive parent-child
    relationships, 63
Conscience, 32
Conscience development, 32
Conscientiousness, 50–53, 57
Constitutional, 46, 48, 113–114, 167
Constraint, 51, 52, 57
Contextual, 31, 33, 43, 167
Continuity, 34, 36, 95, 167
Controlling, 51, 55, 59, 61, 68, 109,
    111
Co-occurrence, 7, 17, 23, 98, 105, 109,
    113, 128, 167
Cooperativeness, 51, 52, 57
Corporal punishment, 70, 71, 95,
    120, 121
Crime, 1, 7, 17, 19, 24, 30, 32–34, 36,
    51, 53, 57, 58, 61, 71, 93, 96, 98,
    100, 101, 103–105, 108, 113, 116,
    135–137, 139
Crime rates, 16, 17, 24
Criminal behaviour, 16, 29, 32, 37,
    42, 70, 76, 94, 98, 100, 101, 103,
    105, 128
Criminal career, 18, 53
Cross-reporting, 136, 137, 168
Culture, 3, 4, 11, 46, 56, 84, 89, 99,
    123, 130–134, 138
Curiosity, 52, 126, 138

Damage, 9, 11, 19
Deceit, 26, 35, 54
Decision-making, 80, 167
Defensive (aggression), 22
Deficient nurturing, 63, 76
Defining, 2, 3, 11, 18–20, 50, 96
Definition, 1–3, 11, 12, 19, 20, 25, 52,
    92, 95, 106, 111, 132
Dehumanization, 87

Delay gratification, 28, 168
Deliberate, 2, 3, 12, 22, 23
Delinquent, 30, 32, 33, 55, 69, 72
Desensitization, 41, 62, 125
Destruction of property, 17, 35, 94, 105
Destructive, 8, 10, 26, 86, 96, 99, 120, 136
Deviance Generalization Hypothesis, 42, 91, 97, 98, 102, 109, 113, 115, 128
Deviant, 2–4, 12, 42, 653, 75, 77, 92, 113, 131, 132
Diagnostic and Statistical Manual of Mental Disorders, 10, 11, 15, 26, 29, 35, 38, 95, 99, 168
Diagnostic criteria/criterion, 3, 6, 34, 36, 99, 128
Discipline, 43, 52, 59, 63, 70, 76, 131
Discrimination, 4, 86, 135, 139
Disengagement, 13, 41, 78, 85–87, 89, 124, 125
Disordered, 3, 45, 47, 51, 57, 99, 113, 132
Disorganised attachment, 63–67, 166
Displace, 87
Displaced aggression, 13, 14, 113, 123, 168
Disposition/dispositional, 26, 31, 33, 43, 45, 125, 168
Disregard, 19, 34, 35, 86, 87, 102
Disruptive, 16, 69, 75
Distortion, 65, 86, 87, 89
Domestic violence, 72, 73, 101, 102, 109–111, 117–118, 120–122, 128, 136
Drive theory, 40, 168
Drugs, 32, 100, 102
Dysfunctional (family, parenting), 33, 55, 56, 106, 109, 136, 168

Early adulthood, 25, 29, 30, 33, 34, 46, 56, 62, 116
Early starters, 28, 37
Ecological factors, 58, 168
Education, 5, 6, 15, 31, 32, 43, 58, 61
Effect sizes, 60, 61, 168
Effortful control, 28, 49, 56, 80, 89, 126, 130, 138, 169

Egocentricity, 53
Egocentrism, 53, 169
Electroencephalogram, 47, 169
Elementary school years, 28, 29, 35, 74, 80, 84
Emotion(s), 22, 26–28, 37, 52, 53, 63–65, 68, 72, 76, 78–80, 89, 130, 169
Emotional, 8, 14, 21–23, 25, 27, 32, 48, 52, 53, 55–57, 68, 74, 78, 79, 107, 111, 128
Emotional reactivity, 32, 33, 56
Emotional violence, 111
Emotion dysregulation, 79, 80, 169
Emotion processes, 44, 78, 80, 89, 169
Emotion processing, 28, 56, 57, 126, 130
Emotion recognition, 55, 72
Emotion regulation (inc. difficulties), 25, 26, 28, 31, 33, 37, 64, 67, 68, 72, 79, 80, 89, 116, 119, 125–128, 138
Empathic arousal, 32, 169
Empathy, 4, 28, 32, 33, 50, 52, 53, 57, 62, 83, 94, 100, 119, 125–127, 129–131
Empirically supported, 1, 22, 130, 169
Encoding, 80, 169
Enuresis, 7, 9, 107, 169
Environmental factors, 1, 43, 44, 47, 57, 58, 72, 75, 76, 78, 114
Erskine, Lord, 6
Escalate/escalation, 17, 25, 96, 97, 128
Euphemistic language, 86, 88, 169
Excitement seeking, 53
Expectation biases, 78, 82, 83, 124
Experiential factors, 80, 170
Exploration, 126
Exposure, 16, 32, 43, 47, 60–63, 72, 73, 76, 77, 107, 112, 118, 124–125, 131
Externalizing (behaviours), 10, 28, 46, 63, 66–68, 70–73, 80, 97, 99, 132, 138, 170
Extraversion, 52
Eysenck's PEN model, 50–52

Facial display, 27
Failure to plan ahead, 35

Family conflict, 63, 76, 112, 122
Family dysfunction, 33
Family environment, 56, 63, 73, 74, 76, 84, 88, 109, 122, 123, 124
Family instability, 31, 122
Family interaction, 41
Family stressors, 43
Family violence, 95, 98, 109–111, 117, 136
Fearless, 49
Fearlessness, 49, 56
Federal Bureau of Investigation (FBI), 46, 98, 105, 106, 113, 128
Fighting, 10, 29, 34, 69, 86, 96, 99, 120, 121
Fight response, 40, 170
Fire setting, 7, 9, 29, 107, 128
First Strike, 93
Five Factor Model, 50, 52
Flight response, 40, 170
Forceful, 53
Four stages of cruelty, 6
Frustration, 27, 37, 40, 48, 49, 56, 68, 107
Frustration-aggression hypothesis, 40

Gangs, 30, 32, 37, 41, 75, 84
Gender, 45, 47, 170
General Aggression Model, 39, 41
Generalizability, 94, 138, 170
Genetic, 21, 43, 45, 68, 71, 72, 80, 170
Goal blocking, 40
Grandiosity, 53, 107, 170
Guilt, 32, 33, 50, 53, 54, 100, 126
Guns, 16, 60

Habituation, 41, 47, 170
Harm, 2, 3, 11, 12, 14, 19, 20, 23, 24, 27, 52, 84, 85, 87, 110, 111, 114, 134, 136
Harsh, 59, 63, 68, 79, 80, 120, 123
Health care workers/professionals, 4, 97, 134
Heart rates, 47
Heightened arousal, 65, 166
Heightened expectation, 56
Heritability, 45, 56, 170
Heritable, 45
Heterogeneity, 19, 42, 97, 106, 169

Histrionic PD, 101, 102, 115
Hogarth, William, 6
Homicide (rates), 7, 16, 24, 33, 46, 58, 86, 94, 95, 104, 116
Hostile
   affect, 14, 23, 24
   aggression, 21–24, 28, 65, 66, 120
   attribution, 65, 83, 124
   feelings, 13
   intent, 83, 84
   intention, 65, 83, 124
   perception, 22, 66, 83, 124
   perception bias, 82, 83, 89, 124, 170
Hostility, 51, 52, 65, 66
Humane, 6, 11, 15, 93, 105, 126
Hyperactive, 49
Hyperactivity, 56, 99, 132, 170

Impulsive, 21–23, 31, 33, 49, 52, 54
Impulsivity, 10, 31, 33, 35, 49, 52, 56, 107, 170
Inadequate monitoring, 63
Inappropriate, 4, 11, 102, 170
Income, 43, 58
Inconsistency, 68, 125
Indirect (abuse), 20, 71
Individual difference(s), 19, 22, 27, 35, 36, 41, 43–45, 48, 50, 51, 56, 58, 64
Industrialized, 16, 24, 58
Infancy, 25–27, 37, 49, 65–67
Inhumane, 2, 11, 52, 87
Institutionalized, 94, 96, 97, 100, 102, 103, 107
Instrumental, 2, 14, 21–24, 26–28
Insults, 28, 59
Intelligence, 31, 33, 36, 54, 120
Intent (to harm), 20, 27
Intentional/intentionally, 2, 11, 20, 28, 103, 166
Interaction(s), 9, 26, 31, 41, 43, 47, 64, 69, 72, 79, 81, 122, 133
Intergenerational, 16, 170
Internal working models, 64, 171
Interpersonal domain, 53
Interpersonal skills, 27, 126, 130, 138
Interpretation(s), 39, 42, 81, 82
Irresponsibility, 35
Irritability, 29, 35, 49, 56, 107, 171
Italian, 112, 131

Justify, 3, 86, 107
Juvenile(s), 7, 16, 107, 128
  crimes, 36, 58
  psychopathology, 54, 57

Kant, Immanuel, 5
Knowledge structures, 41, 48, 78,
  80–82, 84, 124, 171

Law(s), 4, 6, 35, 58, 108, 129, 130,
  132, 135–139
Legal, 3, 12, 120, 130, 135
Levinson, B.M., 8
Life course persistent, 31, 37, 46, 49,
  96, 97, 128
Link(s), the, 9, 48, 61, 71, 98, 136, 137
Locke, John, 5, 11, 15
Longitudinal, 17, 18, 30, 33, 35, 36,
  46, 48, 49, 60, 61, 65–67, 70, 75,
  80, 83, 84, 94, 171
Long-term stability, 35, 36
Loud noises, 40
Low income, 43, 58
Lying, 10, 17, 28, 35, 105, 120

Macro, 75, 76, 114, 171
Manipulation, 35, 169
Manipulative, 53, 54, 102
Marital
  changes, 31
  cohesiveness, 33
  conflict, 31, 43
  rape, 111
Mass
  media, 41, 84, 123
  murder/murderer, 106, 108, 128
Maternal, 12, 31, 66, 68, 107
Mead, Margaret, 6, 8, 15
Media (violence), 60, 62, 76, 89, 124,
  128
Mental scripts, 25
Micro, 43, 76, 78, 88, 114, 171
Minimizing, 87, 89
Mobility, 32, 58, 76, 107
Model, 16, 39–42, 73, 79, 84, 107,
  118–120, 123
Montaigne, 5
Moral, 3, 13, 41, 52, 56, 60, 78, 85–87,
  89, 125, 134, 135, 137

Multidimensional, 14, 15, 23
Murder, 6, 9, 12, 21, 108
Murderer, 7, 47, 93, 106

Narcissism, 50, 54, 57
Negative
  affect, 40, 68, 72, 122, 171
  emotion, 27, 52, 64, 68, 72, 79, 165,
  171
  emotionality, 51, 52, 57
Neglectful, 63, 122
Neighbourhood, 32, 43, 58, 76, 84,
  88, 89, 171
Neo-association theory, 39–41
Neural pathways, 80, 124
Neuroticism, 52, 56
Non-
  compliant, 26, 65, 68
  conformity, 19
Normative belief, 88–90, 95, 119, 124
Norm-violating behaviour, 26
Novelty seeking, 51, 52, 57
Nurturance, 122

Objectivity, 3
Observation (inc. Observational
  learning), 27, 41, 69, 84, 85, 89,
  92, 120, 123, 171
Obsessive-Compulsive PD, 101, 102,
  115
Openness to Experience, 52
Opportunity, 59, 60, 97
Oppositional behaviour, 26, 29, 68, 69
Oppositional Defiant Disorder, 26
Outcome efficacy, 84, 171
Overcrowding, 43

Parental
  abuse, 71, 76, 121
  disciplinary practices, 63, 76
  emotional unavailability, 66
  negativity, 63
  supervision, 59
  violence, 73
  warmth, 67, 68, 70, 80
Parent-child attachment, 59, 65, 76
Parenting, 9, 31, 33, 43, 55, 56, 59,
  61–63, 67–71, 76, 78–80, 88, 114,
  119, 123, 128, 166

Paternal, 68, 95, 107, 120, 121, 122
Pathological gambling, 101, 102, 115
Pathology, 18, 73, 171
Pathway, 18, 29–31, 33, 41, 63, 73, 75, 80, 82, 88, 113, 118, 122, 124
Peer, 27–32, 34, 36, 37, 41, 43, 55, 72–75, 77, 84, 87–89, 106, 112, 129, 134, 135, 137
Perceptions, 11, 61, 65, 78, 80, 82, 84, 88, 89, 124, 128, 129, 134, 135, 137
Perceptual schemas/biases, 22, 80, 81, 88, 89, 124, 171
Persistent (cruelty to animals), 9, 10, 20, 29, 31, 37, 46, 49, 52, 53, 96, 97, 102, 126, 128, 167
Personality, 10, 26, 43, 45, 49–53, 56–58, 81, 94, 172
Perspective-taking, 28, 126
Pet-facilitated psychotherapy, 8
Philosophical, 3, 20
Physiological, 40, 62, 125, 165, 167
Pleasure, 13, 21, 22, 35, 108, 131
Policy, 129, 136, 138
Porphyry, 5
Positive emotion, 52, 68, 79
Poverty, 59, 76
Power assertive strategies, 63, 70
Predisposition, 43, 48, 57, 172
Premeditated, 21–23, 26, 106
Preschool (inc. preschooler), 27–29, 35, 46, 48, 65, 66
Prevalence, 4, 25, 30, 33–35, 37, 46, 58, 70, 101, 110, 112, 116, 120, 172
Preventative, 18, 97, 134, 172
Primary goal, 13, 21
Problem behaviours, 7, 31, 43, 46, 65, 67, 68, 97, 99
Problem solving, 83, 88
Property, 11, 17, 19, 20, 32, 35, 94, 98, 103–105, 135
Prosecution, 136
Prospective, 51, 172
Provocation, 21, 22, 26, 27, 40, 59, 74, 76, 83, 123, 168
Psychopath, 53
Psychopathology, 18, 32, 126, 172
Psychopathy, 43, 45, 50, 51, 53, 54–57, 96, 98–100, 113, 115, 126, 134

Psychoticism, 51, 52, 57
Punishment, 13, 31, 55, 56, 69–71, 76, 95, 120–122, 131, 166
Punitive, 68, 69, 79, 121
Pythagoras, 5

Recurrent (cruelty to animals), 9, 63, 92–95, 98
Rejection, 9, 26, 67, 74, 75, 84
Relapsed lifestyle, 37
Remorse, 35, 53
Residential mobility, 58, 76, 107
Response access, 80
Responsiveness, 65, 67, 166
Retaliatory, 28
Retrospective, 7, 92, 94, 95, 109, 111, 112, 120–122, 173
Reward(s), 14, 21, 23, 28, 32, 52, 55, 75, 85, 167, 168, 173
Rights, 2, 4, 10, 19, 34, 102, 167, 168
Risky family, 56, 63, 76, 122, 124, 125
Rules, 10, 19, 166, 167

Safety, 35, 109
Satisfaction, 21
Schemas, 39, 43, 82–84, 88
School shootings, 93, 106, 107
Scientific
    attention, 93, 106, 107
    community, 135
Scripts, 3, 134, 135
Secure attachment, 64, 66, 79, 165
Selective attention, 80, 82
Self-
    consciousness, 50, 57
    control, 28, 101
    derived standards, 85
    directedness, 52, 57
    efficacy, 84
    esteem (inflated), 26, 75, 85
    exoneration, 87
    image, 60, 89
    judgements, 41
    monitoring, 85
    perception, 100
    reacting, 85
    regulatory inhibitions, 59
    reinforcement, 85

reports(s), 13, 33, 61, 94, 95, 102, 112, 115, 121, 132
transcendence, 52
worth, 60, 100
Sensation-seeking, 19, 49
Sense of maturity, 31
(In)Sensitivity, 26, 56, 67, 134
Sentience, 34
Serial
    killer(s), 93, 106, 108
    murderer(s), 93
Sex, 4, 29, 34, 37, 45–47, 61, 71, 94, 95, 101, 103, 105, 108, 116
Sex differences, 27, 30, 33, 35–37, 44–47, 56, 115–116, 131, 133
Sexual abuse, 71, 107, 111, 114, 120–122
Sexual homicide, 116
Sexual violence, 110
Shallow emotions, 53
Single
    mother, 31
    parent(hood), 43, 58, 76
Situational(ly), 14, 32, 39, 40, 42, 43, 48, 59, 60, 76, 79
Social
    cognitive deficits, 25
    cognitive models, 39–41
    environment, 43, 81
    institution, 19
    learning processes, 41, 56, 57
    learning signals, 27, 82
    learning support systems, 59
    learning theory, 41, 84, 85, 123, 173
    norms, 35, 88
    processes, 41, 56, 57
    scripts, 82
Social class/strata/status, 61, 62, 85
Socialization, 27, 28, 47, 56, 83, 173
Socially unacceptable, 2, 4, 11, 12, 19
Societal
    attitudes, 3
    laws, 58, 168
    level, 2, 4
    norms, 10, 19, 28, 167, 173
Society, 3, 4, 6, 12, 16
Socio-economic status, 19, 24, 86, 135, 171, 173

Stable, 16, 17, 19, 24, 26, 34, 45, 46, 57, 66
    career, 37
    family life, 33
Stability, 25, 27, 35–39
Stalking, 111
Stimulus-response associations, 48
Strategies, 28, 31, 63, 70, 78, 84, 88, 97, 122, 134, 135, 138, 172
Stressful life events, 59
Sub-culture, 41, 84, 123
Superficial charm, 53, 100
Symbolic modelling, 41

Televised violence, 41
Tellegen's three factor model, 50–52
Temperament, 25, 26, 32, 33, 43, 45, 48–50, 56, 58, 67, 122, 124, 169–171
Temper tantrums, 29
Testosterone, 47
Thoughtless, 21, 22
Threat(s), 7, 22, 28, 64, 68, 85, 136, 137
Toddlerhood, 26, 27, 37
Trajectory/trajectories, 17–19, 24, 31–33, 37–39, 57, 76, 96, 97, 122, 127, 128, 173
Triad, the, 7
Trigger, 27, 123, 168

Unconscious, 41, 173
Unemployment, 58
Uninhibited, 26, 32, 49, 87
Unplanned, 21
Unsupportive, 63, 122

Vandalism, 11, 29
Verbal
    abuse, 20, 112, 168
    deficits, 56
Vicarious learning, 118, 171
Vicarious reinforcement, 85, 173
Victim, 4, 7, 14, 20, 21, 23, 85, 87–89, 106, 107, 111, 112–113, 114, 122, 135–137, 139
Victimization, 112, 113
Video games, 60, 62, 107

Violate, 10, 19, 102, 167, 168
Violence Graduation Hypothesis, 42, 91, 93, 94–97, 102, 105, 106, 127
Violent offenders, 7, 36
Violent television (TV), 60, 62
Vulnerability, 14, 32, 53

Warmth, 53, 57, 63, 67, 68, 70, 76, 80, 122
Weapon, 29, 60
Western (inc. culture), 130, 131, 134
Witness, 62, 73, 74, 82, 99, 111, 112, 114–119, 121, 124, 128, 131, 132